I0468865

Two Firefighter Deaths in Auto Parts Store Fire
Chesapeake, Virginia

Investigated by: J. Gordon Routley
Jeff Stern

This is Report 087 of the Major Fires Investigation Project conducted by Varley-Campbell and Associates, Inc./TriData Corporation under contract EMW-94-4423 to the United States Fire Administration, Federal Emergency Management Agency.

Homeland Security

Department of Homeland Security
United States Fire Administration
National Fire Data Center

U.S. Fire Administration Fire Investigations Program

The U.S. Fire Administration develops reports on selected major fires throughout the country. The fires usually involve multiple deaths or a large loss of property. But the primary criterion for deciding to do a report is whether it will result in significant "lessons learned." In some cases these lessons bring to light new knowledge about fire--the effect of building construction or contents, human behavior in fire, etc. In other cases, the lessons are not new but are serious enough to highlight once again, with yet another fire tragedy report. In some cases, special reports are developed to discuss events, drills, or new technologies which are of interest to the fire service.

The reports are sent to fire magazines and are distributed at National and Regional fire meetings. The International Association of Fire Chiefs assists the USFA in disseminating the findings throughout the fire service. On a continuing basis the reports are available on request from the USFA; announcements of their availability are published widely in fire journals and newsletters.

This body of work provides detailed information on the nature of the fire problem for policymakers who must decide on allocations of resources between fire and other pressing problems, and within the fire service to improve codes and code enforcement, training, public fire education, building technology, and other related areas.

The Fire Administration, which has no regulatory authority, sends an experienced fire investigator into a community after a major incident only after having conferred with the local fire authorities to insure that the assistance and presence of the USFA would be supportive and would in no way interfere with any review of the incident they are themselves conducting. The intent is not to arrive during the event or even immediately after, but rather after the dust settles, so that a complete and objective review of all the important aspects of the incident can be made. Local authorities review the USFA's report while it is in draft. The USFA investigator or team is available to local authorities should they wish to request technical assistance for their own investigation.

The U.S. Fire Administration greatly appreciates the cooperation received from Division Chief Thomas H. Cooke.

For additional copies of this report write to the U.S. Fire Administration, 16825 South Seton Avenue, Emmitsburg, Maryland 21727. The report is available on the Administration's Web site at http://www.usfa.dhs.gov/

U.S. Fire Administration

Mission Statement

As an entity of the Department of Homeland Security, the mission of the USFA is to reduce life and economic losses due to fire and related emergencies, through leadership, advocacy, coordination, and support. We serve the Nation independently, in coordination with other Federal agencies, and in partnership with fire protection and emergency service communities. With a commitment to excellence, we provide public education, training, technology, and data initiatives.

 Homeland Security

TABLE OF CONTENTS

Two Firefighter Deaths In Auto Parts Store Fire
Chesapeake, Virginia
(March 18, 1996)

Investigated by: J. Gordon Routley
Jeff Stern

Local Contact: Division Chief Thomas H. Cooke
Fire Marshal
Chesapeake Fire Department
304 Albemarle Drive
Chesapeake, Virginia 23320
(757) 382-6165

OVERVIEW

On March 18, 1996, two firefighters were killed in Chesapeake, Virginia when they became trapped by a rapidly spreading fire in an auto parts store and pre-engineered wood truss roof assembly collapsed on them. The cause of the fire was an electrical short created when a power company truck working in the rear of the building drove away with its boom in an elevated position, accidentally pulling an electrical feed line from the main breaker panel at the rear of the store. Post-incident investigations indicate that the electrical fault may have sparked multiple points of fire origin throughout the roof structure of the building, due to improperly grounded wiring.

This is another incident illustrating the rapid failure of lightweight construction systems when key support components are involved in a fire. It points out the importance of pre-fire planning and accurate size up by fire companies to determine the risk factors associated with a fire in this type of construction. Lessons regarding importance of initial company actions, constant re-evaluation of action plans, strong command and coordination of units on the fireground, and recognition of signs of impending structural failure were also reinforced.

This incident was investigated by the City of Chesapeake Fire Department Fire Marshal's Office. Incident investigations and analysis were also conducted by the National Fire Protection Association, the Virginia Power Company, and the Virginia State Occupational Safety and Health Administration.

The Chesapeake Fire Department serves a rapidly growing area of 353 square miles in Tidewater, Virginia, just west of Virginia Beach and south of the City of Norfolk. The department has 319 career firefighters and EMS personnel that provide service from 14 fire stations. The estimated population in the City of Chesapeake is 170,000. The initiatives taken by the Chesapeake Fire Department as a result of this incident are listed in Appendix C to this report.

1

SUMMARY OF KEY ISSUES

Issue	Comment
Staffing	The first alarm response provided a small attack force with limited capabilities. The full response brought only 10 personnel.
Size-up	The first arriving company officer was not able to determine the location and extent of the hidden fire.
Pre-fire plan Information	This complex required a pre-fire plan due to the complex arrangement, multiple occupancies, mixed construction, lack of fixed protection, limited access and difficult water supply problems. The first-due company did carry a pre-fire plan that showed the layout of the shopping center and the floor plan for the auto parts store, but the pre-fire plan was not referenced by the crew during the fire.
Delayed response	The first arriving company was on the scene alone for several minutes with only 3 personnel. The back-up companies had long response times. The lack of evidence of a working fire prompted the initial incident commander to return some of the responding units, resulting in even longer response times.--
Water Supply	The first-in company did not establish a water supply. This required the second engine company to be committed to this task.
Incident Command	The battalion chief was faced with a complicated and rapidly changing situation. He was not able to effectively transfer command from the initial officer and direct the operations of widely separated units.
Operational risk management	The officers involved in the initial part of the operation had to make critical risk management decisions with limited information.
Accountability	Accountability for the personnel operating in the hazardous area was not established prior to the structural collapse. As the situation became critical, no one realized that a crew was still inside the building.
Rapid intervention crew	Additional crews did not arrive in time to assist the crew that was in trouble inside the building.
Radio communications	The lack of a clear radio channel for fireground communications caused serious problems with command and control of the incident, including the failure to maintain communications with the crew inside and the failure to hear their request for assistance.
Lightweight construction	The roof collapsed quickly and with very little warning. This should be anticipated with a lightweight wood truss roof assembly. This hazard was not recognized by the crews on the scene.

BUILDING DESCRIPTION

Construction and History

The fire occurred in a modern, lightweight construction building that was added to an existing strip mall in 1984. The older mall on exposure side four was separated from the fire building by a masonry fire wall and was constructed with masonry walls and a steel bar-joist roof structure. The exposures on side two consisted of additional stores that were similar in construction to the auto parts store. There were no exposures on sides one and three (Figure 1).

The auto parts store was constructed with two masonry exterior walls and two wood frame exterior walls, with a lightweight wood truss roof assembly. It was approximately 120 feet deep and 50 feet wide, providing about 6,000 square feet of open display and storage space. The roof assembly was a pre-engineered lightweight wood truss assembled from 2 x 6 top and bottom chords, with 2 x 4 web members held together with metal gusset plates. There were no interior bearing walls or supports for the roof structure. At one end, the trusses were supported by a wood plate that was bolted to a metal beam[1]. The other end rested on top of the concrete block wall. Each truss was separated by 24 inches and they were covered with 1/2 inch CDX plywood sheathing under a two-ply rubber membrane. A drywall ceiling was attached to the underside of the trusses, creating a truss void space (truss loft) 24 to 36 inches above the ceiling. A drywall divider was located in the middle of the truss void as a draft stop. The roof had a slight pitch (Figure 2).

Three air handling units were on the roof of the building, with an estimated combined weight of 3,000 pounds. It is not known when these units were installed and they may have represented an unanticipated dead load on the roof assembly. There was no indication that the trusses had been reinforced to support the extra weight of these units.

The original truss roof structure collapsed during the construction of the building, injuring three workers. Most of the trusses were damaged and had to be replaced at the time.

The fire building was occupied by Advance Auto Parts, a chain distributor of automobile parts and lubricants. The store was designed with an open retail area containing display racks for goods. A long counter ran from front to back behind which was shelving for additional auto parts. Waste oil and batteries were kept in a rear storage area separated from the front of the store by a drywall wall. The southwest corner of the building contained employee restrooms which had a small water heater located in the ceiling space just above them. (Figure 3).

The main entrance to the store was through two large glass doors at the front of the building. A delivery and service entrance was located in the rear and a 40 foot trailer was parked behind the building and used for additional storage.

---

[1] Post incident investigations by the Chesapeake building inspector revealed that these ends may not have been properly secured to the beam.

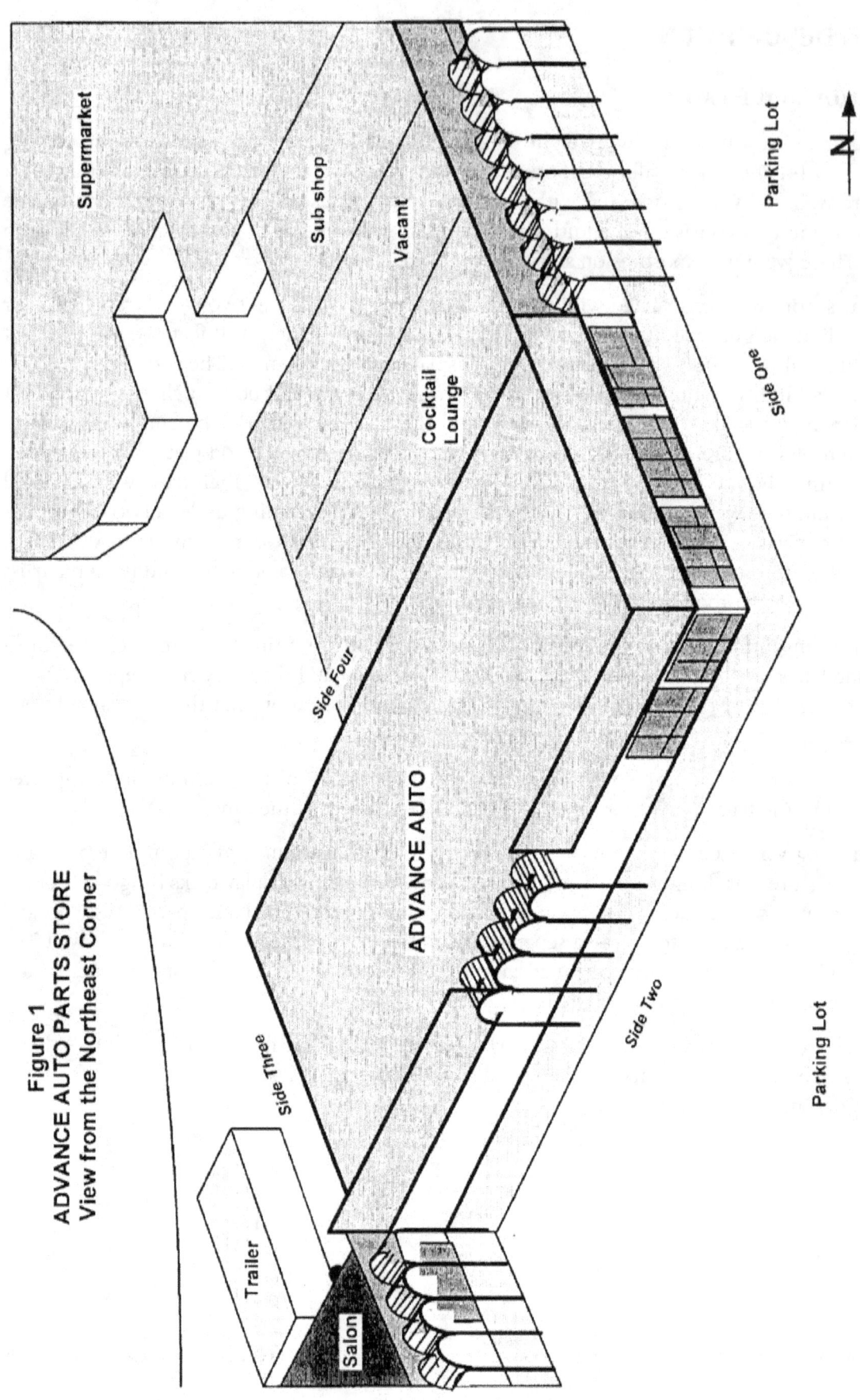

Figure 1
ADVANCE AUTO PARTS STORE
View from the Northeast Corner

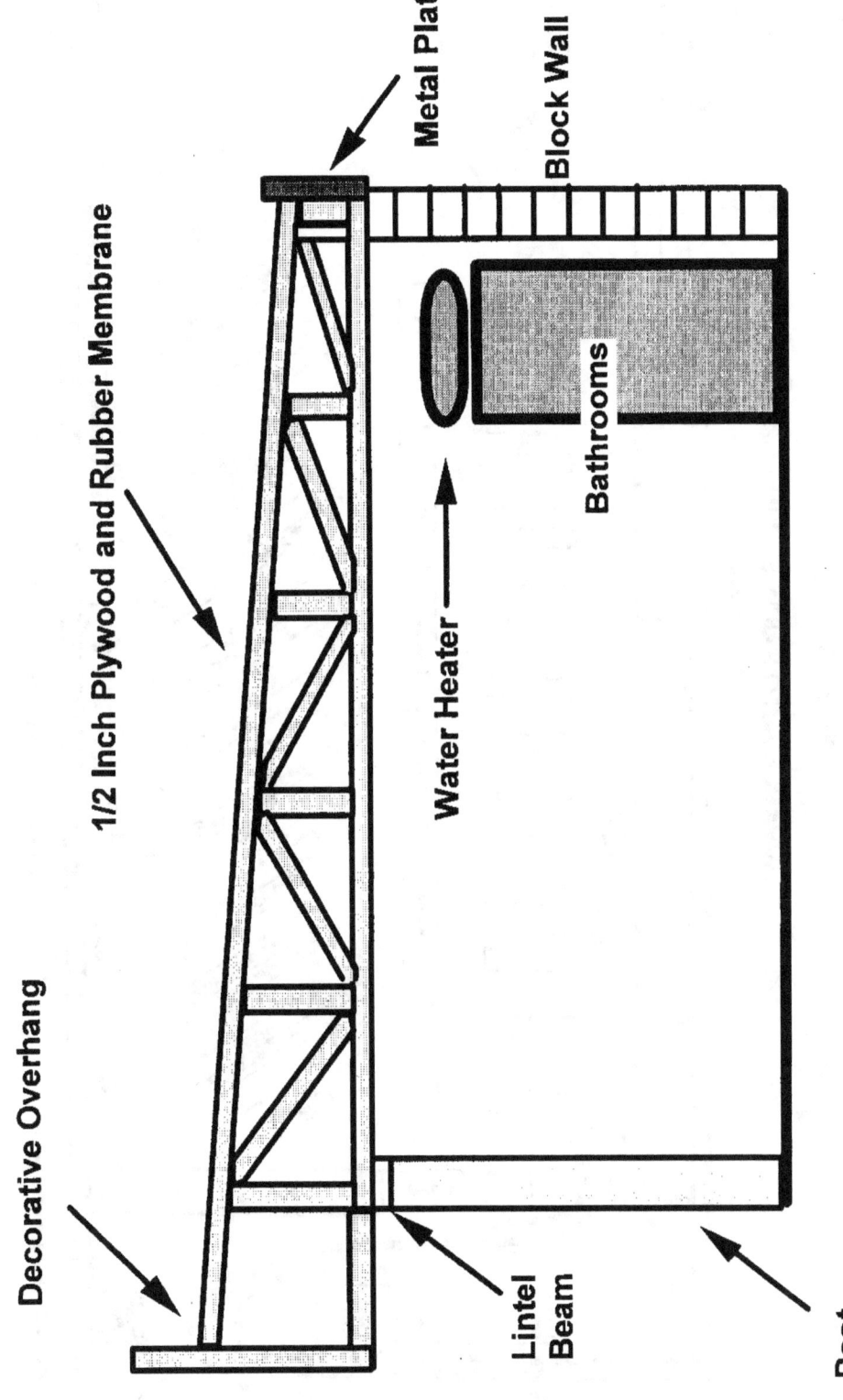

Figure 2
ADVANCE AUTO PARTS STORE
Dissection of the truss roof as viewed from front to back
of the building.

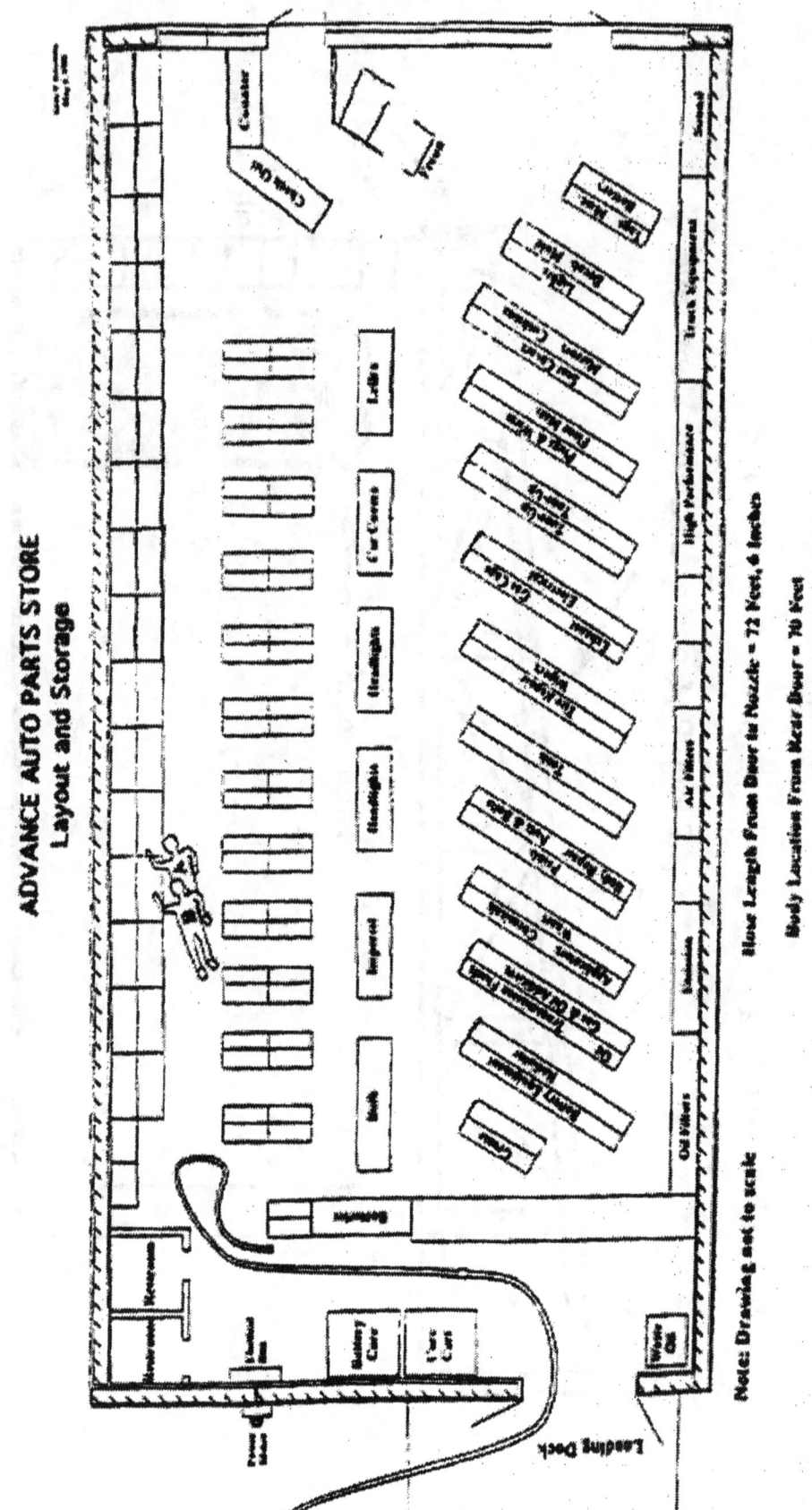

Figure 3

ADVANCE AUTO PARTS STORE
Layout and Storage

THE FIRE

Cause

At approximately 11:00 a.m. on March 18, 1996, a power company employee set up a service truck at the rear of the Indian River Shopping Center in Chesapeake, Virginia. The worker was going to disconnect the electrical power to a customer who had not paid an electric bill. The customer, a cocktail lounge and bar, was located adjacent to Advance Auto Parts. In preparing to disconnect service, the power company worker elevated the articulating boom on his truck to roof level. Faced with the immediate loss of power, an employee of the lounge paid the electric bill while the power company employee was beginning work, and went to the back of the store to show the receipt. A stamped receipt indicates the bill was paid at 11:16 a.m. at a supermarket also located in the shopping center. The power company employee, working from the bucket of the articulating boom, lowered the boom and verified the receipt. Although the bucket had been lowered, the hinged elbow of the articulating boom remained elevated. The employee then radioed his supervisor from the cab of his truck, and received instructions not to disconnect power.

The power company employee then attempted to drive the service truck away, forgetting to secure the boom, which snagged on a power line feeding the meter at the rear of the Advance Auto Parts store. This caused a phase-to-phase and phase-to-ground arcing fault at the store's electrical meter, starting the fire. The power company employee immediately stopped, exited his truck, and cut the remaining power connections to the meter at the rear of Advance Auto Parts.

Initial Actions Prior to Calling 9-1-1

After cutting the power line to the building, the power company employee removed the meter, noticed smoke coming from the meter base, notified his office and requested that another power company crew and a supervisor come and assist him. An employee of the Advance Auto Parts store came to the rear of the building and met the power company employee, telling him that the store had lost electrical power and that a fire was being extinguished inside the building. Another Advance Auto Parts employee discharged a dry chemical fire extinguisher on the spot fire that had started near the hot water heater above the employee restrooms. All believed the fire had been extinguished at this time.

At 11:29 a.m., the Chesapeake Fire and Police Emergency Operations Center received a 9-1-1 call from Advance Auto Parts reporting a problem with the fuse box in the store. The Chesapeake Fire Department was dispatched to a report of a fuse box sparking at 4345 Indian River Road at the Advance Auto Parts store.

Emergency Response

Initial response consisted of two engines, a ladder company, and a battalion chief, for a total of 10 personnel. Engine 3 was the first due arriving company, responding from quarters. Engine 1 and Ladder 2 also responded. Battalion 1 was dispatched as the command officer, but requested that Battalion 2 cover the assignment since he was out of position. Battalion 2 acknowledged the request, and he responded with the first alarm companies.

Engine 3's crew consisted of three personnel: a driver/pump operator; Firefighter Specialist John Hudgins, serving as Acting Lieutenant for the shift; and Firefighter Specialist Frank Young, detailed to the station for the day, was riding in the jump seat. Engine 3 was responding in a reserve engine that had a 500 gallon water tank.

Initial Size-Up and Company Actions

At approximately 11:35 a.m., about five and a half minutes after dispatch, Engine 3 arrived on the scene at the front of the strip mall. Hudgins reported "a single-story commercial structure, nothing showing from the front. Engine 3 is in command."

Engine 3 took a position in front of the Advance Auto Parts Store. Hudgins and Young entered the structure from the front of the building to investigate. Conditions were clear in the store, and there was no visible smoke or flames showing. They discovered light smoke near the electrical panel in the rear of the building, and radioed to Battalion 2 that they had a fire and were checking for extension. Action Lieutenant Hudgins then radioed for Engine 3's driver to reposition to the rear of the building.

Hudgins then radioed to Battalion 2, who had not yet arrived on the scene, that Engine 3 and Ladder 2 could handle the incident. Battalion 2 and Engine 1, the second due engine company, both went in service.

Original Assignment Dispatched

About one minute later, the crew from Engine 3 discovered additional fire in the building and requested the original assignment continue to the scene. Battalion 2 and Engine 1 both acknowledged and started back to the call.

The driver/pump operator of Engine 3 observed light smoke coming from the roof-line on side two as he repositioned to the alley on side three of the store. Hudgins and Young met him at the rear of the building, where heavier smoke was showing from the edge of the roof. The firefighters pulled two 1 3/4 inch attack lines off Engine 3. Firefighter Young used the first attack line to sweep the rear of the building, where flames had started to come from the roof-line at the top of the rear wall. Acting Lieutenant Hudgins donned his SCBA and entered the rear of the building with the second attack line. Engine 3's driver directed firefighter Young to assist Acting Lieutenant Hudgins, who was inside the building. Young closed the nozzle on his attack line, placed it on the ground, and followed the first attack line into the rear of the building.

Battalion Chief's Arrival and Second Alarm

Approximately eleven minutes after the time of dispatch, Battalion 2 marked on the scene, observing light gray smoke at the decorative arches in front of the building and a light haze of smoke inside the structure. Less than one minute later, Battalion 2 radioed for a second alarm, reporting "flames showing," and requested police assistance with evacuating exposure occupancies. Two additional engines were dispatched for the second alarm. The battalion chief then proceeded to evacuate the exposures on sides two and four of the auto parts store (Figure 4).

Hudgins radioed to Battalion 2 that they had a fire in the ceiling and needed an additional crew to don SCBAs and come inside with pike poles to pull ceiling. Battalion 2 acknowledged the transmission and radioed that he had "a fire showing – building fire showing from outside."

Engine 3's driver radioed to Battalion 2 that he needed the second due engine company to lay a supply line to him at the rear of the building. Battalion 2 radioed to Engine 1 to confirm that they had understood the request.

Engine 1 and Ladder 2 Arrive, Third Alarm Called

Engine 1 marked on the scene approximately thirteen minutes after dispatch. As they passed by the front, a firefighter on Engine 1 noticed smoke and some fire in the front of the store, rolling down to the floor, and observed a strong draft pulling air into the building. Engine 1 then proceeded to the rear to reverse lay a supply line from Engine 3 and establish the water supply.

Once in the rear, the firefighter and the officer from Engine 1 observed one attack line from Engine 3 that stretched into the back door of the Advance Auto Parts Store. Heavy smoke was visible inside and fire was coming from the roof in the rear of the store. The officer instructed Engine 3's driver to set up a 2 1/2 inch attack line. In the time it took to prepare this line, the firefighter from Engine 1 reported that the rear entrance to the building had become fully involved with fire, and the attack line into the rear door had burned through. Engine 3 ran out of water at this time and was unable to charge the 2 1/2 inch attack line. Engine 1's officer attempted to contact Battalion 2 by radio, but was unable to get through.

Simultaneously, Ladder 2 was arriving in the front of the building and taking a position on side two of the store. While the diver was setting up the aerial to go to the roof, the officer and firefighter from Ladder 2 were preparing to enter the front of the building to assist Engine 3's crew, but the officer noted heavy smoke inside the store and a large volume of air being drawn in through the front doors.

He determined that it was unsafe for his crew to enter the building and advised his driver to set up for defensive operations. Ladder 2's officer then met with Battalion 2 in front of the building and suggested that a third alarm might be needed due to the growing size of the fire.

About eighteen minutes after dispatch, Battalion 2 reported "a working commercial building fire fully involved." Within the next minute, he transmitted a request for a third alarm. At this time, Engine 1 was still in the process of laying five inch hose from Engine 3 to a fire hydrant located at the far corner of Indian River Road and MacDonald Road.

Engine 3 Reports They Are Trapped, Roof Collapses.

At approximately 11:49 a.m., almost 20 minutes after the initial dispatch time, Hudgins radioed that he and Young could not get out of the building. Battalion 2 radioed back that he could not understand their transmission. Hudgins then radioed that they needed someone to come to the front of the building and get them out. Again unable to understand their transmission, Battalion 2 radioed for any unit on the fireground to advise him if they heard the message that was transmitted. Engine 4 responded that they were unable to copy the transmission. Engine 14 then marked the scene and was instructed by Battalion 2 to lay a supply line to the front of the building. Battalion 1, enroute to the fire on the second alarm, radioed to Battalion 2 that it sounded like someone was trapped inside. Battalion 3, also enroute, radioed that he would be on the scene momentarily and would assist.

At this time, Ladder 2's crew was setting the outriggers and preparing to elevate their aerial ladder for defensive operations. In the short time it took to accomplish the stabilization of the ladder truck, the front of the store became fully involved, the building contents ignited, and the roof collapsed.

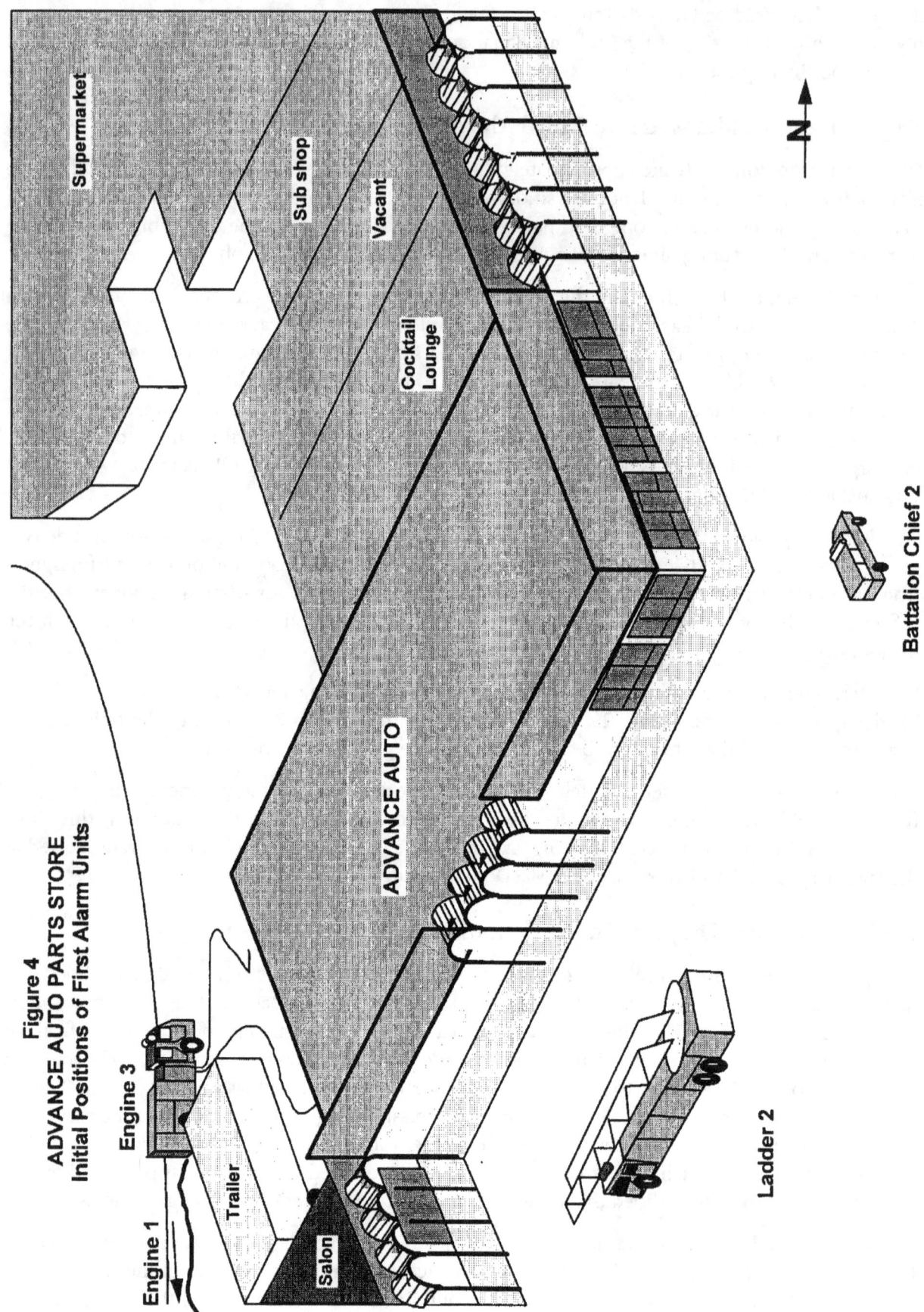

Figure 4
ADVANCE AUTO PARTS STORE
Initial Positions of First Alarm Units

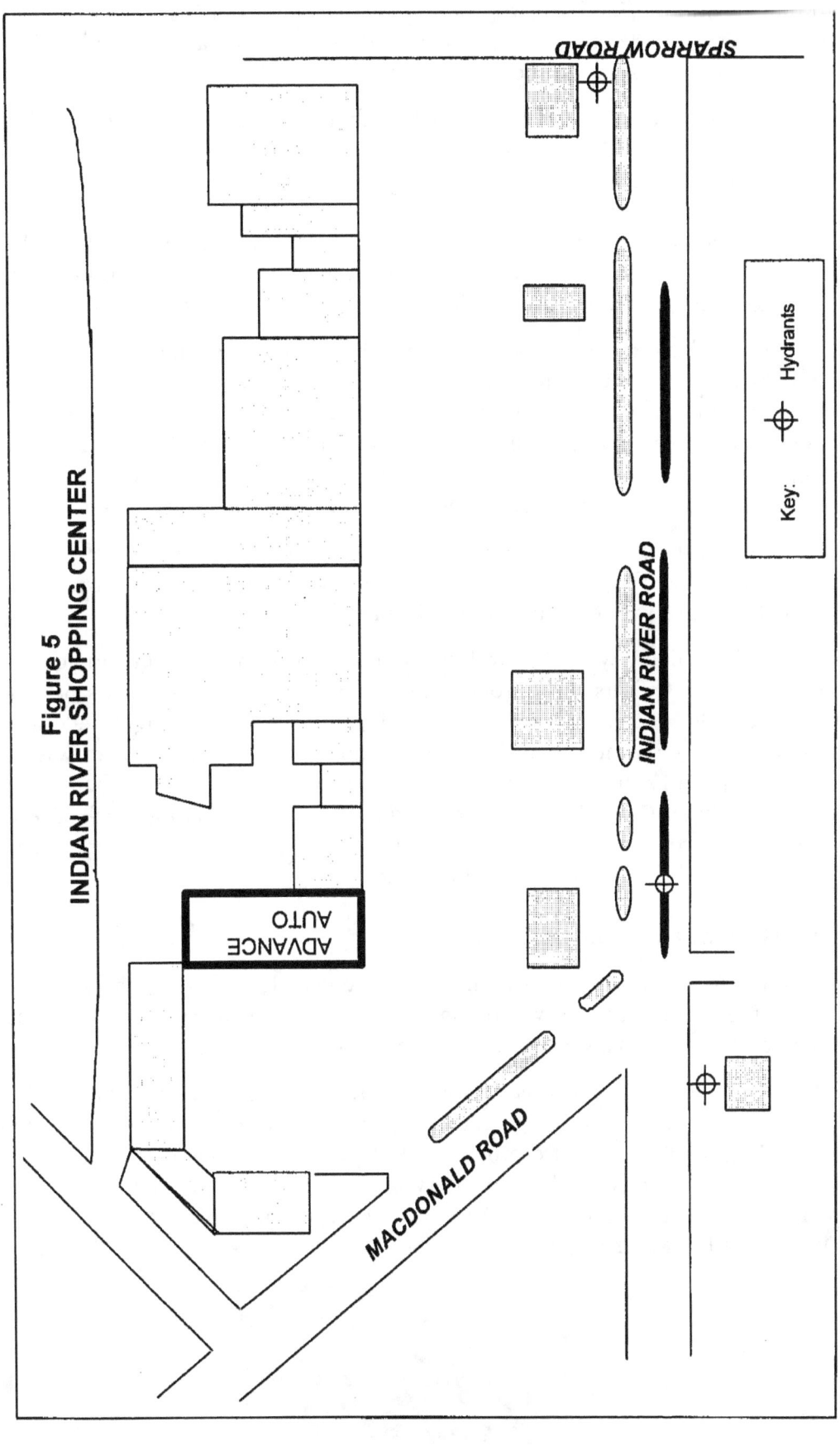

Figure 5
INDIAN RIVER SHOPPING CENTER

Due to the radiant heat, Ladder 2 was forced to retrace their outriggers and reposition to a safer defensive position on side one of the structure, and set up the aerial again. Ladder 2's crew did not hear Engine 3's transmission that they were trapped.

Simultaneously, Engine 1 ran out of supply line about 200 feet short of the hydrant. Engine 2, responding on the second alarm, picked up the hydrant that Engine 1 was attempting to reach and laid a supply line to one side. The driver of Engine 1 attempted to contact his officer by radio to advise that he could not reach the hydrant, but could not get through due to heavy radio traffic. He parked the engine in the roadway, donned his SCBA, and went to the rear of the building to report to his captain and rejoin his crew. Battalion 3 arrived on side one about this time and radioed for all companies to switch to channel two, an alternate fireground tactical frequency.

Driven by the northerly wind and the draft created by the burning contents of the structure, the fire at the rear had grown in such intensity that personnel were forced to move Engine 3. Assisted by employees of the power company, Engine 3 was moved back away from the rear of the building. At 11:55 a.m., about 26 minutes after dispatch, the Captain of Engine 1, with his crew at the rear of the building, confirmed to Battalion 2 that "I got men on the inside from Engine 3, and the lines have been burned. I do not know their status, and we still have no water to go in after them."

Battalion 3 met with Battalion 2 and discussed that they may have lost a crew inside. Battalion 3 assumed command and Battalion 2 went to the rear of the building to coordinate rescue efforts. There, Battalion 2 met with the Captain from Engine 1.

By this time, the building was fully involved and no rescue efforts could be mounted until the fire was knocked down. Officers at the front and the rear attempted to conduct a personnel accountability report (PAR) to determine who was missing and where they might be located. An engine company responding on mutual aid from the Virginia Beach Fire Department was flagged down, connected to Engine 1's supply line, and completed the water supply to a hydrant behind the shopping center within the City of Virginia Beach. Engine 3 was forced to move back once again, and the supply line was disconnected from Engine 3 and used to supply water to Engine 4, a telesquirt that was positioned for defensive operations at the rear.

Extinguishment and Body Recovery

The fire spread to the attic of the exposures on side two and was held in check by the fire wall on side four of the building. The fire was brought under control as the contents of the auto parts store burned off and several aerial streams were put into operation.

After the fire was extinguished, a search for the missing firefighters was initiated. After the bodies of the firefighters were located, they were packaged and removed from the fire building by members of the Virginia Beach Fire Department, and transferred by members of the Chesapeake Fire Department to medic units. The body recovery was supervised by the Chesapeake Fire Department Fire Marshal's Office and documented. An investigation was immediately started by the Chesapeake Fire Department Fire Marshal.

TIMELINE OF EVENTS

(Constructed from Emergency Communications Center Audio Tapes and Witness Interviews)

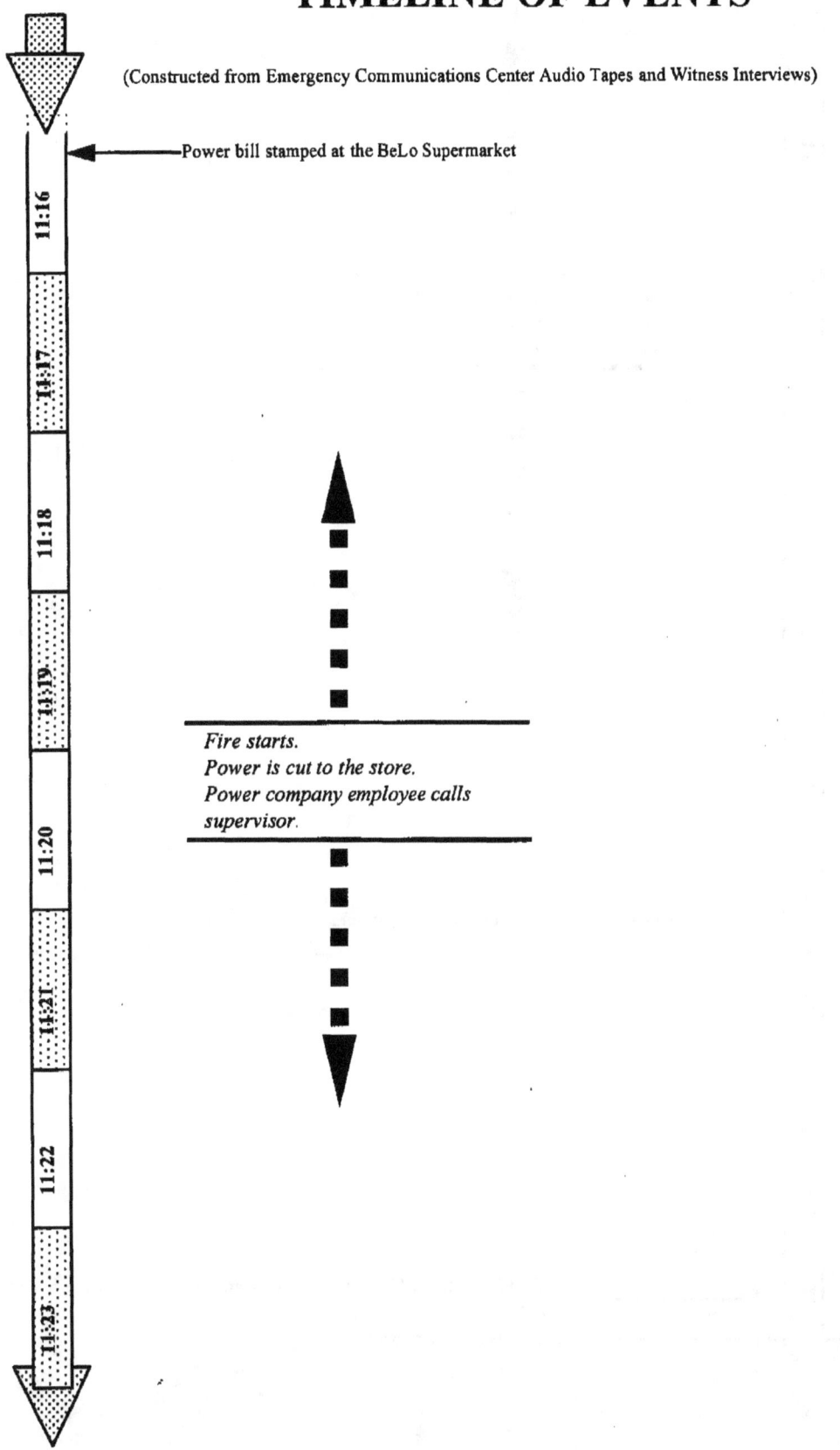

Power bill stamped at the BeLo Supermarket

11:16

11:17

11:18

11:19

11:20

11:21

11:22

11:23

Fire starts.
Power is cut to the store.
Power company employee calls
supervisor.

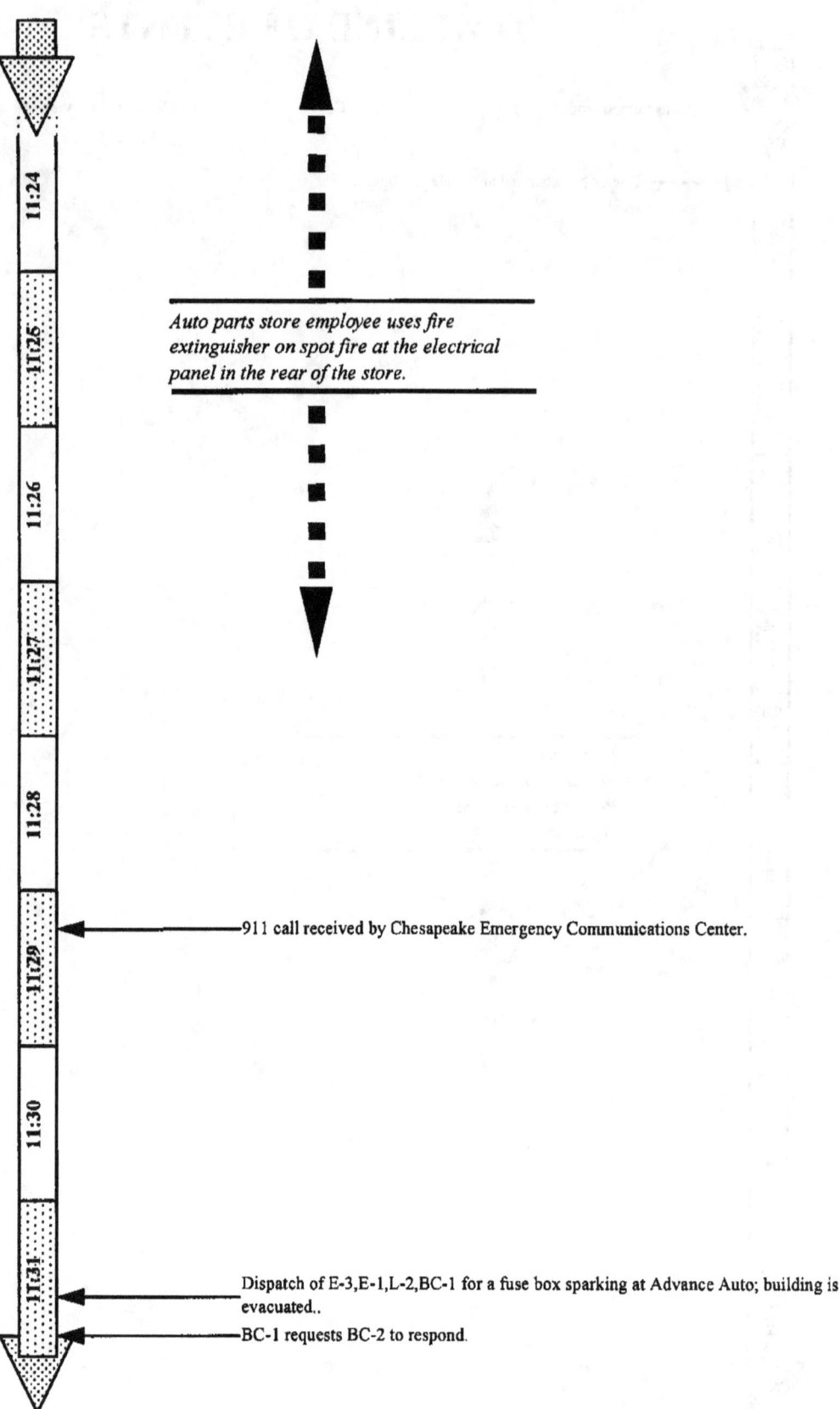

11:24

11:25

11:26

11:27

11:28

11:29

11:30

11:31

Auto parts store employee uses fire extinguisher on spot fire at the electrical panel in the rear of the store.

911 call received by Chesapeake Emergency Communications Center.

Dispatch of E-3,E-1,L-2,BC-1 for a fuse box sparking at Advance Auto; building is evacuated..

BC-1 requests BC-2 to respond.

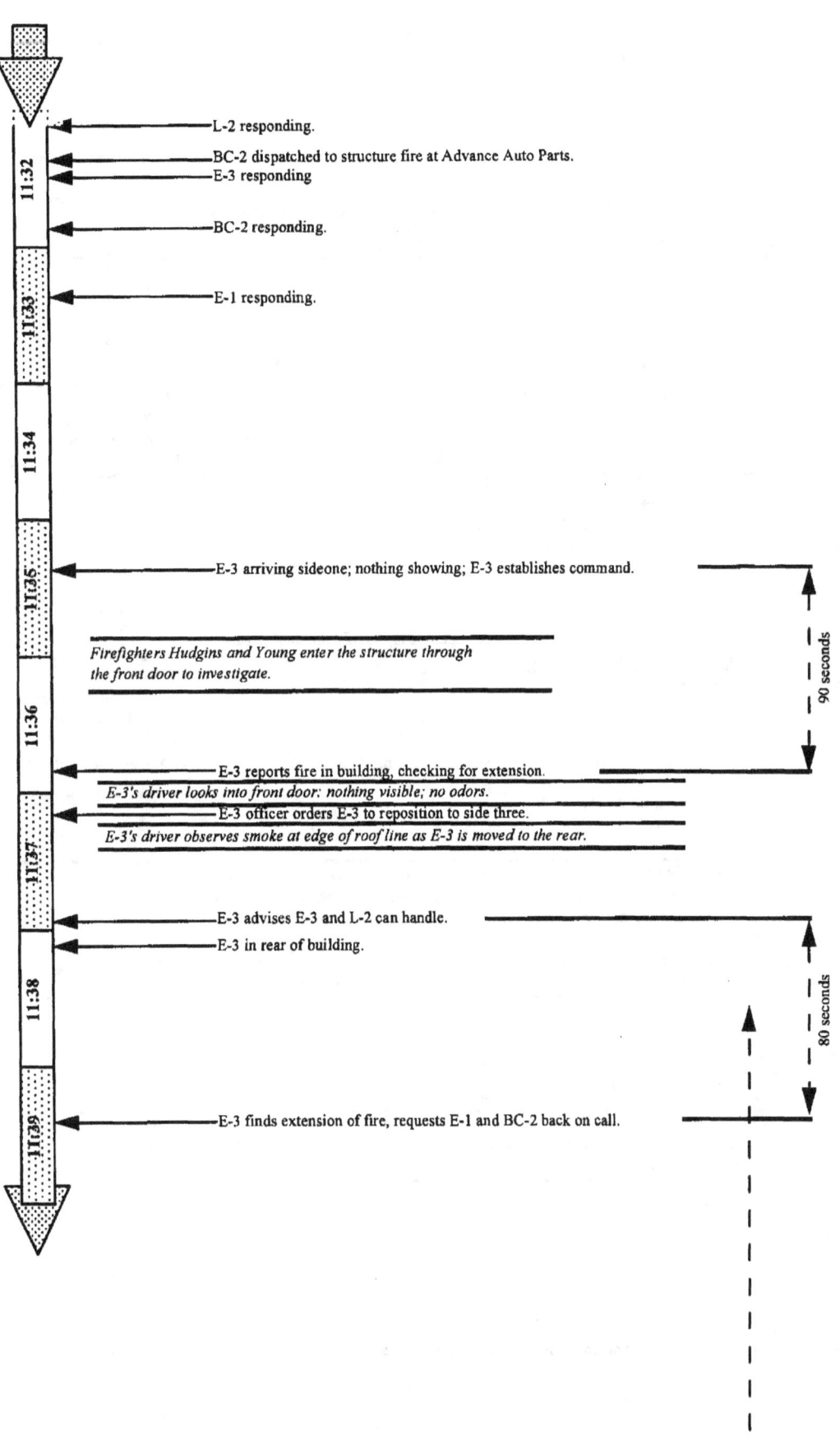

L-2 responding.

BC-2 dispatched to structure fire at Advance Auto Parts.

E-3 responding

BC-2 responding.

E-1 responding.

E-3 arriving sideone; nothing showing; E-3 establishes command.

Firefighters Hudgins and Young enter the structure through the front door to investigate.

E-3 reports fire in building, checking for extension.

E-3's driver looks into front door: nothing visible; no odors.

E-3 officer orders E-3 to reposition to side three.

E-3's driver observes smoke at edge of roof line as E-3 is moved to the rear.

E-3 advises E-3 and L-2 can handle.

E-3 in rear of building.

E-3 finds extension of fire, requests E-1 and BC-2 back on call.

90 seconds

80 seconds

11:32

11:33

11:34

11:35

11:36

11:37

11:38

11:39

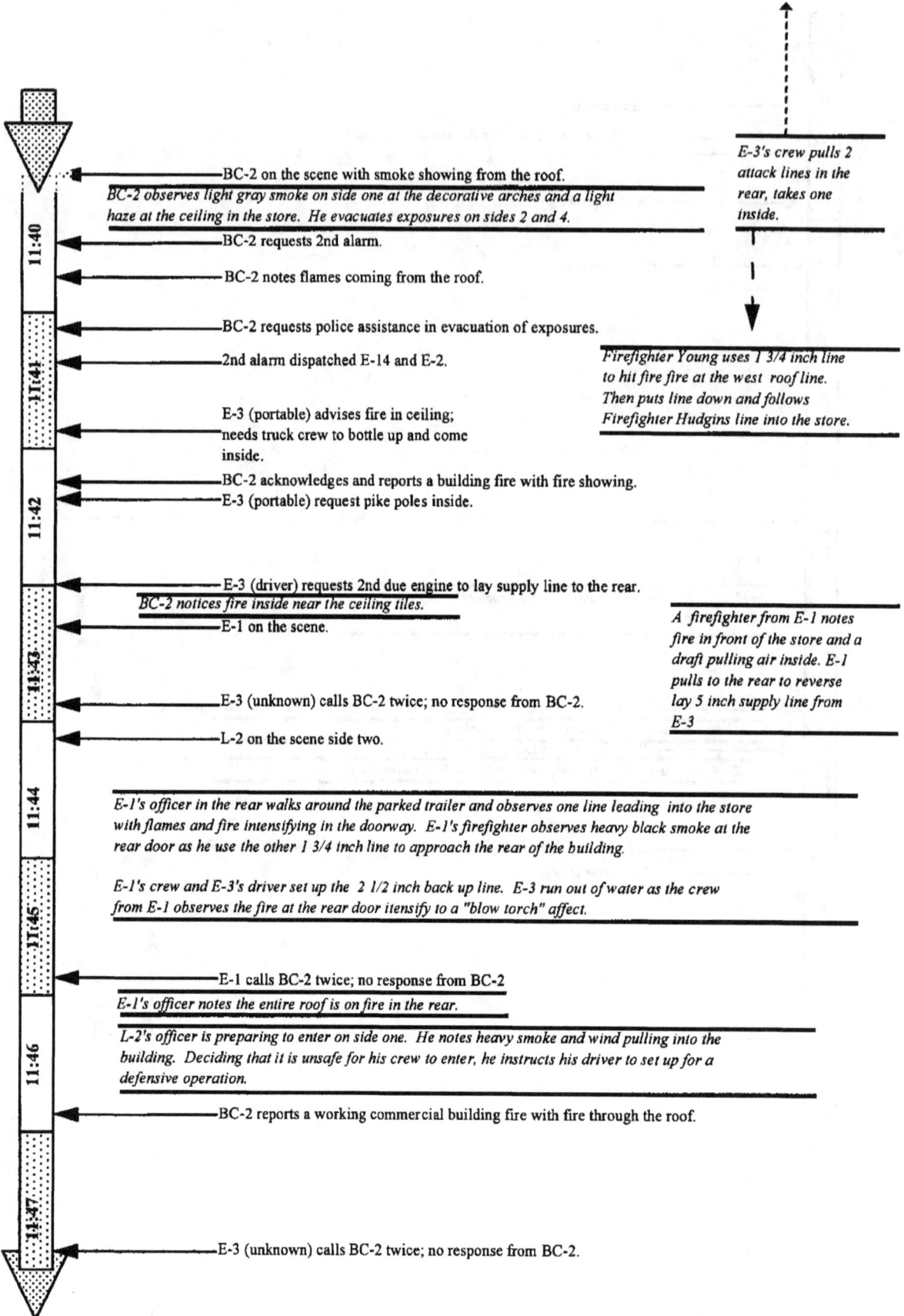

BC-2 on the scene with smoke showing from the roof.

BC-2 observes light gray smoke on side one at the decorative arches and a light haze at the ceiling in the store. He evacuates exposures on sides 2 and 4.

BC-2 requests 2nd alarm.

BC-2 notes flames coming from the roof.

BC-2 requests police assistance in evacuation of exposures.

2nd alarm dispatched E-14 and E-2.

E-3 (portable) advises fire in ceiling; needs truck crew to bottle up and come inside.

BC-2 acknowledges and reports a building fire with fire showing.

E-3 (portable) request pike poles inside.

E-3 (driver) requests 2nd due engine to lay supply line to the rear.

BC-2 notices fire inside near the ceiling tiles.

E-1 on the scene.

E-3 (unknown) calls BC-2 twice; no response from BC-2.

L-2 on the scene side two.

E-1's officer in the rear walks around the parked trailer and observes one line leading into the store with flames and fire intensifying in the doorway. E-1's firefighter observes heavy black smoke at the rear door as he use the other 1 3/4 inch line to approach the rear of the building.

E-1's crew and E-3's driver set up the 2 1/2 inch back up line. E-3 run out of water as the crew from E-1 observes the fire at the rear door itensify to a "blow torch" affect.

E-1 calls BC-2 twice; no response from BC-2.

E-1's officer notes the entire roof is on fire in the rear.

L-2's officer is preparing to enter on side one. He notes heavy smoke and wind pulling into the building. Deciding that it is unsafe for his crew to enter, he instructs his driver to set up for a defensive operation.

BC-2 reports a working commercial building fire with fire through the roof.

E-3 (unknown) calls BC-2 twice; no response from BC-2.

Timeline: 11:40, 11:41, 11:42, 11:43, 11:44, 11:45, 11:46, 11:47

E-3's crew pulls 2 attack lines in the rear, takes one inside.

Firefighter Young uses 1 3/4 inch line to hit fire fire at the west roof line. Then puts line down and follows Firefighter Hudgins line into the store.

A firefighter from E-1 notes fire in front of the store and a draft pulling air inside. E-1 pulls to the rear to reverse lay 5 inch supply line from E-3

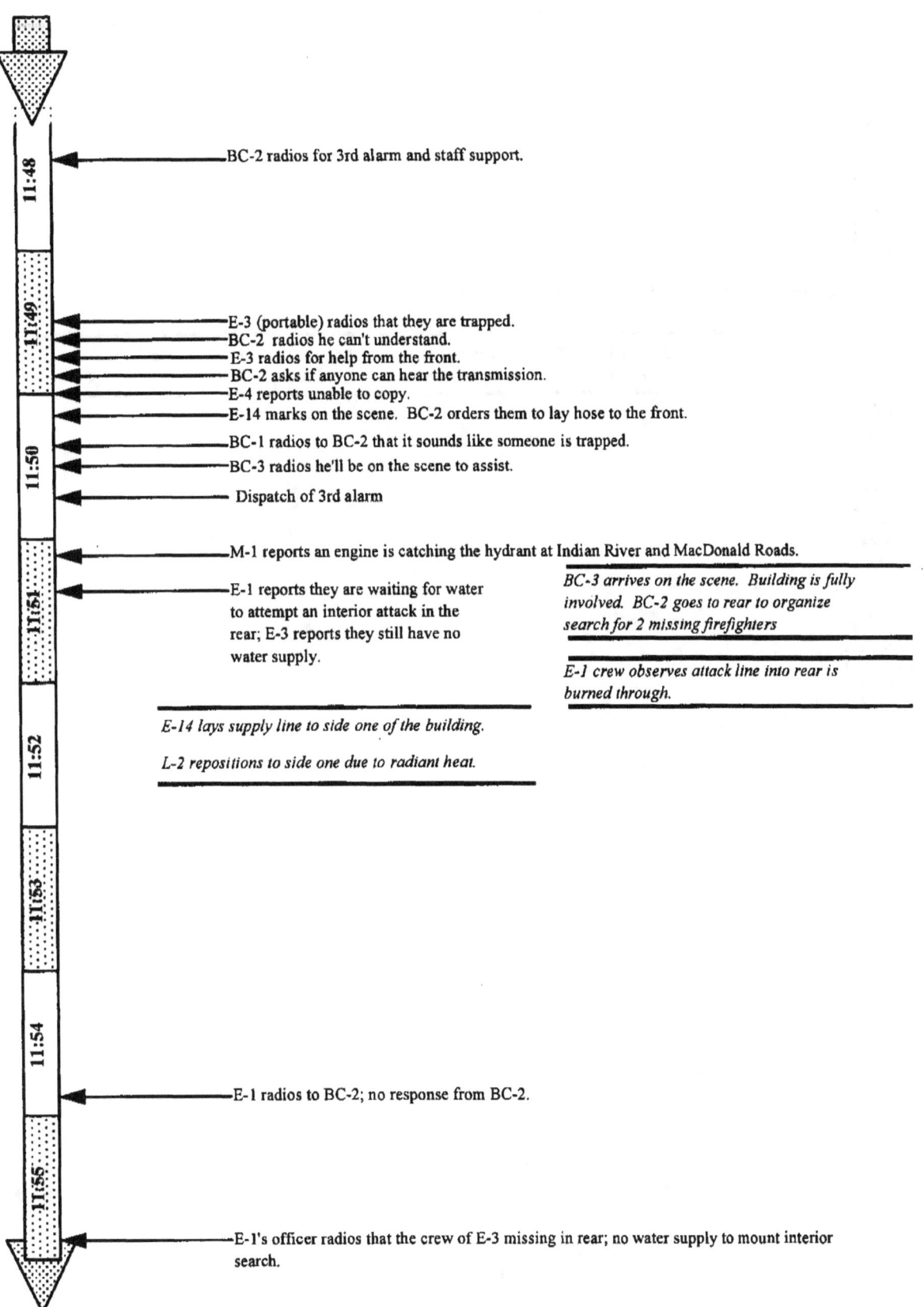

11:48 — BC-2 radios for 3rd alarm and staff support.

11:49 — E-3 (portable) radios that they are trapped.
— BC-2 radios he can't understand.
— E-3 radios for help from the front.
— BC-2 asks if anyone can hear the transmission.
— E-4 reports unable to copy.
— E-14 marks on the scene. BC-2 orders them to lay hose to the front.

11:50 — BC-1 radios to BC-2 that it sounds like someone is trapped.
— BC-3 radios he'll be on the scene to assist.
— Dispatch of 3rd alarm

11:51 — M-1 reports an engine is catching the hydrant at Indian River and MacDonald Roads.

— E-1 reports they are waiting for water to attempt an interior attack in the rear; E-3 reports they still have no water supply.

BC-3 arrives on the scene. Building is fully involved. BC-2 goes to rear to organize search for 2 missing firefighters

E-1 crew observes attack line into rear is burned through.

11:52 — *E-14 lays supply line to side one of the building.*

L-2 repositions to side one due to radiant heat.

11:53

11:54 — E-1 radios to BC-2; no response from BC-2.

11:55

— E-1's officer radios that the crew of E-3 missing in rear; no water supply to mount interior search.

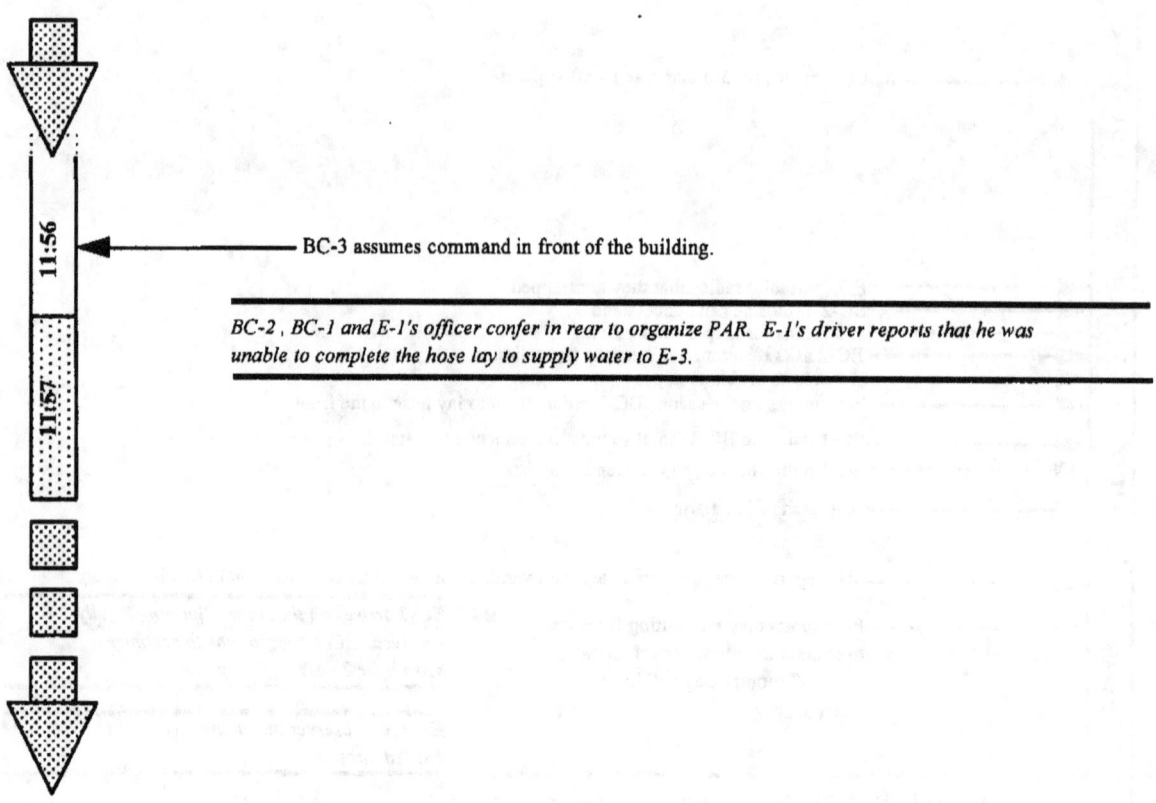

11:56 — BC-3 assumes command in front of the building.

BC-2 , BC-1 and E-1's officer confer in rear to organize PAR. E-1's driver reports that he was unable to complete the hose lay to supply water to E-3.

ANALYSIS

Fire Cause and Flame Spread

The fire was caused by the electrical short created when the power company truck struck the power line to the building. Investigation by the City of Chesapeake Electrical Inspector after the fire revealed that the meter contained wiring that appeared to have been tampered with and did not comply with the electrical code. Several connections at the meter had been doubled-lugged, connecting multiple wires to single terminals. Additional investigation by Virginia Power revealed that the building may have been improperly grounded, leading to numerous hot connections when the short circuit occurred. The main fuse did not trip at the breaker panel and the wiring on all three air handling units had been fused. This probably resulted in the ignition of multiple spot fires in the truss loft above the store.

It appears that the fires in the truss loft were still relatively minor when Engine 3 arrived, but the fire spread rapidly throughout the space due to the light wood construction. The wind drawn from the open doors at the front of the building also promoted rapid fire growth. This would have created a tremendous hidden fire in the wood truss loft area despite clear conditions inside the structure. Reports of heavy smoke and fire conditions on the roof at the same time Engine 3's crew was calling for pike poles and personnel to come inside are indications towards this scenario.

The interior of the auto parts store contained racks of auto parts and supplies, including oil, lubricants, rubber and plastic parts. The contents were packed closely together and stored in tall racks near the ceiling. Once the fire had broken through the ceiling in the rear of the building, these contents would have quickly reached their ignition temperatures, creating flashover conditions in the rear of the store as the fire progressed, trapping the firefighters and forcing them to seek an exit at the front of the store.

Roof Collapse

The collapse of the pre-engineered truss roof occurred approximately 21 minutes after the time of dispatch, and within 35 minutes of the initial accident that caused the electrical short. The structure appears to have collapsed within 10 to 12 minutes after the truss space became heavily involved. The collapse of similar truss assemblies under fire conditions within this time period has been well documented.

Post-incident investigations indicate that this truss assembly may have been weakened by deficiencies in the connection of the trusses to the beam on the east side of the building. Also, the dead load of the three air conditioning units may have contributed to the rapid failure of the roof.

Reports from firefighters on the scene indicate that a partial failure of the truss assembly may have occurred in the rear of the building, followed shortly by the failure of the entire roof assembly. It is possible that the crew of Engine 3 was trapped by the partial collapse of the roof in the rear, or by the collapse of racks containing auto parts in the building, or by the rapid spread of the fire and smoke which had broken through the ceiling. It is also possible that a combination of these events occurred simultaneously. The failure of the entire roof assembly and complete involvement of the interior of the building with fire took place within one minute after the firefighters radioed for help, before any reaction to assist them could take place.

Cause and Nature of Firefighter Deaths

The cause of the deaths of Firefighters John Hudgins and Frank Young can be attributed to becoming caught and trapped by rapidly spreading fire and falling debris, followed immediately by the collapse of the trussed roof. Autopsies determined the nature of their deaths as burns from direct flame exposure. The autopsies indicated that Firefighter Specialist Young, who was found with his mask on, had a blood carboxyhemoglobin level of 11 percent. Firefighter Specialist Hudgins, found without a mask on, had a blood carboxyhemoglobin level of 25 percent.

Fire Operations

Initial Response- The first alarm assignment was overwhelmed by the situation, the circumstances, and the unusual sequence of events that occurred at this incident. It is evident that a larger force would have been needed to initiate an effective offensive or defensive operation for a working fire in a 6,000 square foot commercial occupancy, with attached exposures on two sides, with or without the unusual complications. The response of two engine companies, one ladder company and a battalion chief, provided a total of only 10 personnel on the initial assignment. The individual companies, which responded with three person crews, had limited capabilities to perform tasks independently.

This incident generated only a single call to 9-1-1 reporting an electrical problem. In most communities only a small percentage of potential structure fire responses actually turn out to be working fires and most of those incidents can be identified from the information that is received from callers. It is a common practice for many fire departments to send a limited initial response to possible structure fires and to dispatch additional companies if the first arriving units report a working fire. Many fire departments also dispatch additional companies if multiple calls are received or the information indicates a probable working fire.

In this case, there was no indication of a working structure fire when Engine 3 arrived. Response times for the second arriving engine and the ladder company were relatively long due to the location of the incident, which was in a remote point within the city limits of Chesapeake. The second due engine was further delayed by two to three minutes as a result of being canceled, then redispatched, As a result, Engine 3 was on the scene with only three personnel for eight minutes before any other companies arrived.

The response times for multiple alarm companies responding from other Chesapeake stations were also long, due to the location of the fire. The first of these companies was just arriving when the roof collapsed. After the first due engine, the next closest companies were located in the City of Virginia Beach; however, these companies were requested on mutual aid only after the second alarm.

Offensive Attack - Engine 3 had been on the scene for approximately four minutes before the crew realized that there was a working fire above the ceiling and the officer reinitiated the response of the second due engine and the battalion chief. At this point, the fire was spreading rapidly within the truss loft. The company initiated an interior offensive attack on the fire in the truss space of the auto parts store. The first preconnected line from Engine 3 was operated from the exterior, where flames had broken through the panels that enclosed the trusses. This line was operated for approximately one minute before the second line was advanced into the rear section of the auto parts store.

The battalion chief arrived and assumed command of the incident at about the time the line was being taken inside the building to begin the interior attack. At this point the fire in the truss space was growing rapidly and there were only four personnel at the scene: the officer and firefighter from Engine 3 were operating the line inside the building, the pump operator was outside at the rear, and the battalion chief had assumed command of the incident at the front of the shopping center. The engine officer, who was operating the hoseline inside, could not see how quickly the fire was developing above the ceiling. The pump operator knew that the officer and firefighter were inside, but his concern was directed toward the need for a supply line, because Engine 3 was still operating on tank water. Their reserve apparatus did not have a functioning tank gauge, which made it impossible to determine how much water was left in the tank.

The battalion chief was alone at the front of the complex, where he could see heavy smoke and flames coming from the roof at the rear of the auto parts store. The visible fire conditions prompted him to transmit a second alarm almost immediately after assuming command. He knew that Engine 3 was operating at the rear, but he was not specifically aware of who was inside or what conditions they were encountering at that time. The need to evacuate exposures conflicted with the need for a strong command presence at this critical stage of the incident.

Engine 1 was committed to the rear as soon as it arrived in order to establish the water supply for Engine 3. While the captain and the firefighter from Engine 1 hooked up the supply line to Engine 3, the driver began to lay a 5 inch supply line to a hydrant. Engine 1's officer could see that Engine 3's line went inside the building and recognized the need for a 2 1/2 inch back-up line. By the time the back-up line was ready, flames were issuing from the door where the attack line had entered, but the back-up line could not be operated before the supply line was charged. No one at the rear anticipated or was aware of the problem of Engine 1 running out of hose before reaching a hydrant.

The offensive attack also required support functions, which are usually performed by ladder companies. The multiple support functions that are required for a fire in a building of this size would normally require at least two ladder companies; however, only one ladder company was dispatched on the initial assignment.

Before Ladder 2 arrived, Engine 3 had requested assistance inside the store to open ceilings and expose the fire. Upon arrival, the officer and firefighter from Ladder 2 prepared to enter, but by the time they reached the front door, the smoke and fire conditions inside the auto parts store made entry impossible. The driver who remained outside, began to set up the aerial ladder to reach the roof; however, no additional personnel were available to provide ventilation or perform other outside functions at that time. Conditions were deteriorating quickly at that point, making it necessary to reposition the ladder truck for defensive operations. The officer and firefighter had to assist the driver in moving the vehicle. There was no one immediately available to assist the two firefighters inside the building, and their request for help went unheard by crews on the scene.

Incident Management - The Chesapeake Fire Department has a documented incident command system which is in accordance with National incident management practices and guidelines; however, incident management was disorganized prior to the collapse of the roof. Several factors affected the management of this incident, including the difficulty in establishing water supply, calling for additional resources, and difficulty in communicating with units on the fireground while additional units were dispatched on the same frequency. The rapid fire spread and failure of the trussed roof assembly emphasized the need for command to be established and coordinated early in the incident, and for clear strategic goals to be determined, prioritized and carried out tactically under the supervision of sector commanders.

Operational Risk Management

There are several important issues related to operational risk management at this incident. Every fire officer must be prepared to make these types of critical decisions based on a risk management assessment of each operational situation. When examining the decisions that were made at this incident, it is important to consider the information that was available to the individuals who had to make those decisions at the time.

The first risk management issue relates to the decision to initiate an interior attack; the officer in charge of Engine 3 had to decide on an appropriate action for the situation, based on his assessment of the risks that were involved. The second risk management issue relates to the failure to evacuate the building before the interior attack crew became trapped.

Was it appropriate for Engine 3 to initiate an interior attack with the number of personnel on the scene?

The analysis of this incident has caused a significant discussion over the number of personnel who are required to be on the scene of a structure fire to initiate an interior attack. NFPA 1500, the Standard for a Fire Department Occupational Safety and Health Program, 1992 edition, states that a minimum of two personnel must enter and work together as a team inside a hazardous area and that at least one individual must remain outside and must maintain an awareness of the location and function of the entry team members. This requirement could be satisfied with the three crew members on Engine 3, the officer and firefighter inside and the pump operator outside.

A tentative interim amendment (TIA) to NFPA 1500, adopted in 1993, added a requirement that a minimum of 4 members must be assembled at the scene before initiating an interior attack at a working structure fire, which is define as a fire that requires the use of breathing apparatus and at least a 1 1/2 inch attack line. This would be at least nominally satisfied by addition of the battalion chief, who arrived at about the same time as the attack line was being taken into the building

The Occupational Safety and Health Administration (OSHA) of the U.S. Department of Labor issued an interpretation on May 1, 1995, which is based in part on NFPA 1500 and in part on the OSHA standard for respiratory protection 29 CFR 1910.134. This interpretation has been referred to as the "2-in/2-out rule." It makes reference to the NFPA requirement for a minimum of two personnel to work together inside a hazardous area – using a "buddy system." It also refers to a requirement in the OSHA regulation for "men" to remain outside with the capability of rescuing workers who are using respirators in an atmosphere that is immediately dangerous to life and health (IDLH).

The NFPA and OSHA requirements could be satisfied with four personnel on the scene, if two members remained outside while two entered the hazardous area. To meet all of the requirements the personnel outside would have to account for and maintain communications with the personnel inside, and would have to be prepared to assist or rescue them.

The Federal OSHA interpretation was issued in the form of an enforcement policy directive to agencies that enforce the Federal OSHA regulations. The Virginia Department of Labor and Industry enforces these regulations under a "State plan agreement" with the U.S. department of Labor. At the time of the incident, the Virginia Department of Labor and Industry had not announced its intentions with regard to an enforcement policy for fire departments in Virginia or taken any enforcement actions.

Accountability - The Chesapeake Fire Department had a standard operating procedure and had practiced the use of a personnel accountability system, but it was not implemented until after the roof collapse occurred. The Chesapeake Fire Department uses a system of magnetic helmet shields and plastic name tags. Each firefighter uses a helmet shield for the unit he is assigned to for the shift. The plastic name tags are placed on a common 2 inch by 5 inch Velcro strip or "passport" in each unit, and are then passed on to the sector officers or incident command post at the emergency scene. The system is used by all departments in the Tidewater, Virginia region. The Chesapeake Fire Department's Incident Command System manual also provides for the establishment of an Accountability Officer to administer the Personnel Accountability System (PAS) at emergency incidents.

Since no sector officers were initially assigned to the rear of the building at E-3's location, the passport would have been inaccessible for Battalion 2, at the front of the building. CFD PAS policy states that "passports and tracking of those companies should be delegated to a sector officer as soon as possible." Due to the limited number of personnel on the fireground, nobody was available to be assigned to this rear sector until after the firefighters became trapped. The policy also calls for the dispatcher to use an alert tone and notify the incident commander after every twenty minutes of elapsed time in a working incident. This notification was not made. A personnel accountability report (PAR) is also called for at any report of a missing or trapped firefighter, any change from offensive to defensive modes of operations, or any sudden hazardous event at the incident. An evacuation order was not issued when heavy fire became evident from the roof, nor was a PAR called for during the set up of defensive operations or at the time of the roof collapse.

The CFD's 14 page policy on fireground safety clearly states that "Ground crews must be notified and evacuated from interior positions before ladder pipes go into operation." The policy also calls for the establishment of a Safety Sector. A specific section of this policy is titled "structural collapse" and states that "under fire conditions light weight wood truss and bar joist roof construction can be expected to fail after minimal fire exposure," and goes on to state that "It is the principal command responsibility to continually evaluate and determine if the fire building is tenable for interior operation."

The battalion chief knew from the radio reports that the officer and possibly one or more crew members from Engine 3 were operating inside the building. He could not see the entry point and could not maintain accountability for who was inside or outside at any point in time.

The pump operator knew specifically that the officer and firefighter from Engine 3 had entered the building, but from his vantage point he could not see the door where they had entered and could not be sure if they remained inside or came back outside or if anyone else entered through the back door. His attention was occupied with stretching the back-up line and attempting to obtain a water supply. He later stated that he assumed that the interior crew must have made their way out through the front of the building when the fire conditions became untenable.

It was not realized for several minutes after the roof collapse that the officer and firefighter from Engine 3 were missing. It was impossible to conduct a search until the main body of fire was knocked down.

Communications - Each firefighter on the entry team had a portable radio, but the radio traffic quickly overloaded the single channel that was available for fireground communications. The Chesapeake Fire Department tactical channel was the non-repeated side of the dispatch channel, which was being used to dispatch multiple alarm companies, direct move-up assignments, and for other emergency and non-emergency traffic. This made it impossible to maintain a reliable communications link with the entry team.

The incident commander did not have a clear channel to provide instructions or receive reports from the company officer on the scene. He had left his vehicle and was trying to size-up the situation and evacuate the adjoining occupancies, while using a portable radio to communicate. At least one radio channel should be dedicated to tactical operations at working incidents and someone outside should continuously monitor the tactical radio channel that is used by crews working inside a building or hazardous area.

The radio traffic also made it impossible for the driver of Engine 1 to advise his officer or the incident commander that he had run out of hose. This caused a major delay in completing the supply line for Engine 3, which also contributed to the tragic outcome of the incident.

Rescue Capability - The pump operator and the battalion chief both had protective clothing and breathing apparatus available to enter the building, if they had known the entry team members were in trouble and needed assistance. The pump operator was occupied with the water supply problem, while the battalion chief was busy trying to evacuate the adjoining occupancies and direct the other responding units. However, even if they had been prepared to enter, it is very unlikely that they could have rescued the entry team from the burning interior.

The real rescue capability for the interior crew was Engine 1, which was preparing the 2 1/2 inch back-up line. The rear portion of the auto parts store was already in flames by the time this line was stretched, but it could not be operated until the supply line was charged. The entry team was cut-off from their point of entry and their hoseline came through the area that was involved in the fire. Having run out of water, there was no chance for Engine 1's crew to mount a rescue attempt.

When should the interior crew have been ordered to evacuate the building?

The two firefighters inside the auto parts store may have been the only firefighter's on the scene who could not see that the fire conditions were becoming critical, before they realized that they were trapped. At least eight other firefighters were present outside the building before this occurred and all of them recognized that the magnitude of the fire was increasing quickly. No one warned the entry team to evacuate or heard their request for assistance when they became trapped.

Flames were coming through the roof and the interior was heavily charged with smoke for several minutes before the roof collapsed. The critical risk created by the fire conditions should have been recognized, even if the personnel on the scene did not know about the lightweight roof construction. In spite of the obvious fire conditions, no one warned the interior crew to evacuate.

The battalion chief knew that Engine 3 was working inside because the officer had asked for a team to come inside with breathing apparatus and pike poles to pull the ceiling. The crew of Ladder 2 was responding to that request and had determined that they could not enter safely through the front. The crew of Engine 1 could not see the line that went through the door and the fire conditions in the rear portion of the store. The pump operator assumed that his fellow crew members had made their way out through the front of the building.

Battalion 2 had reported that the building was becoming fully involved and requested the third alarm before the collapse occurred. Instead of concentrating on command functions, the battalion chief was personally involved in taking action, evacuating exposures on sides two and four.

Lightweight Construction – The critical risk factors of a working fire involving a lightweight truss roof assembly and an occupancy with an unusually high fire load were not recognized. Experience has shown that this type of roof can collapse after as little as 10 to 12 minutes of fire involvement. If the wood truss roof construction had been recognized or even suspected, the interior crew should have been ordered to evacuate the building immediately.

Recognition of the wood roof hazard would have required prior knowledge of the building details or reference to a pre-fire plan, since it could not be determined from visual observation at the scene of the fire. The existing pre-fire incident plan was available in a ring binder on Engine 3, but was not referenced by the crew on the response to the fire.[2]

It appears that by the time the inside crew realized they were in trouble and requested assistance, it would have been too late for anyone to rescue them, even if a rapid intervention crew had been standing-by outside the building. The entire roof collapsed very shortly after their message requesting assistance was transmitted. Their request for a crew to come in from the front of the building suggests that they initially became trapped by the fire which cut-off their path to the rear doors. This may have resulted from a partial roof collapse in the rear section of the building or from a ceiling failure which caused the highly combustible contents at the rear of the auto parts store to become involved in the fire. The major part of the roof collapsed within a minute after they reported that they were trapped and requested assistance.

Safety Equipment

All Chesapeake firefighters wear NFPA compliant turnout gear, including PBI/Kevlar turnout coats, rubber boots, protective hoods, helmets, gloves, and SCBA with PASS devices.

SCBA – The SCBA worn by Hudgins and Young were both NFPA compliant MSA brand Ultralite II models, with 1/2 hour air bottles. The SCBA were sent to NIOSH for testing after the incident.

Turnout Gear The turnout gear worn by Hudgins and Young was damaged by the fire conditions, which far exceeded the protective capabilities of the suits due to direct flame exposure. Parts of the protective ensemble underneath the firefighter's bodies were intact. The remains of the protective clothing were secured by the Chesapeake Fire Marshal and examined after the fire. No defects were discovered.

PASS Devices Hudgins and Young both wore PASS devices. Due to the high heat conditions, these units melted when the firefighters were killed, and it was impossible to determine whether they were in the on, off, or alarm position.

LESSONS LEARNED AND REINFORCED

This fire reinforced many operational issues that are essential to safe and effective firefighting operations. (Specific changes made by the Chesapeake Fire Department in response to this fatal fire are listed in Appendix C.)

---

[2] The pre-fire plan for the building incorrectly identified the wood truss roof structure as steel truss, but still may have provided valuable information and forewarned the firefighters that they were under a trussed roof. During its inquiry, the Chesapeake Fire Department determined that John Hudgins had inspected the store two years earlier as part of an existing CFD company inspection program that requires all companies to inspect every commercial occupancy within their first response each year.

1. **RISK ASSESSMENT is the primary responsibility of the incident commander.**

This incident presented a very high risk to the firefighters who were attempting to make an interior attack. However, the risk factors were not recognized and the interior crew was not directed to abandon the building. Risk assessment should be a continual process, particularly when a situation is changing very quickly.

2. **ACCOUNTABILITY is an essential function of the Incident Command System.**

The location and operation of the initial attack crew was not tracked according to the incident command system that was in effect at the time of the fire. The system must keep track of the location, function, status, and assignment of every individual unit or company operating at the scene of an emergency incident. In order to be effective, the accountability process must be routinely initiated at the beginning of every incident, and updated as the incident progresses and units are reassigned to different tasks.

3. **TACTICAL RADIO CHANNELS are essential for firefighter safety.**

The fireground operations were conducted on the same radio channel as the routine dispatch and transfer of additional units, hampering the fireground communications during the important early stages of the incident. Designated radio channels should be set aside specifically for communications between the incident commander and the units operating at the scene of an incident. The exchange of information, orders, instructions, warnings, and progress reports is essential to support safe and effective operations. Tactical channels should be assigned early and routinely to avoid the confusion that occurs when units that are already working are directed to switch to a different radio channel.

4. **FIRE OPERATIONS must be limited to those functions that can be performed safely with the number of personnel that are available at the scene of an incident.**

The initial response to this incident did not provide enough resources to safely initiate an effective interior attack for the situation that was encountered. The first arriving company initiated interior operations that could not be adequately performed or supported with the limited number of personnel at the scene or responding. The delayed arrival of back-up companies increased the risk exposure of the first due company. The situation called for a more conservative initial attack plan and/or an early retreat when the magnitude of the fire became evident.

5. **WATER SUPPLY is a critical component of a safe and successful operation.**

The failed attempt to establish an adequate and reliable water supply for the interior attack was a critical problem at this incident. This task occupied the second due engine company, which was needed to provide either a back-up hoseline to support the interior attack or a rapid intervention crew.

6. **LIGHTWEIGHT WOOD TRUSS CONSTRUCTION is prone to rapid failure under fire conditions.**

If the construction of the building had been known or recognized, the early failure of the roof structure should have been anticipated and the interior crew should have been withdrawn. This requires pre-fire planning to identify high risk properties and a reliable system to label the building or to inform the responding units of the risk factors of the building. It is usually difficult or impossible to make this determination when the building is burning.

APPENDIX A

Pre-Fire Plans

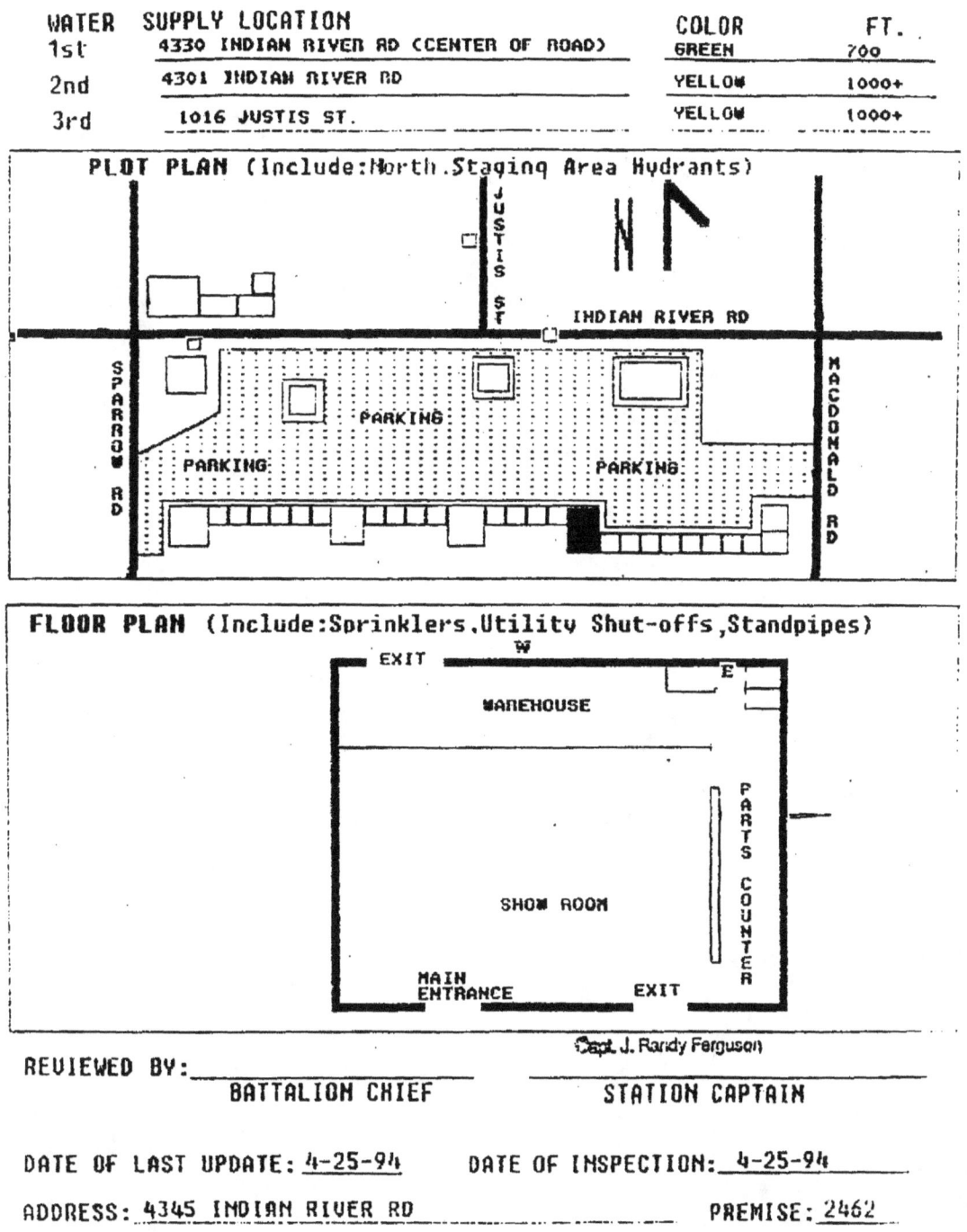

WATER	SUPPLY LOCATION	COLOR	FT.
1st	4330 INDIAN RIVER RD (CENTER OF ROAD)	GREEN	700
2nd	4301 INDIAN RIVER RD	YELLOW	1000+
3rd	1016 JUSTIS ST.	YELLOW	1000+

PLOT PLAN (Include:North.Staging Area Hydrants)

JUSTIS ST

INDIAN RIVER RD

SPARROW RD

PARKING

PARKING

PARKING

MACDONALD RD

FLOOR PLAN (Include:Sprinklers.Utility Shut-offs,Standpipes)

EXIT

WAREHOUSE

E

SHOW ROOM

PARTS COUNTER

MAIN ENTRANCE

EXIT

REVIEWED BY:_____

BATTALION CHIEF

Capt. J. Randy Ferguson

STATION CAPTAIN

DATE OF LAST UPDATE: 4-25-94   DATE OF INSPECTION: 4-25-94

ADDRESS: 4345 INDIAN RIVER RD   PREMISE: 2462

APPENDIX B

Communication Center Transcripts

Page 5
Dispatch Tapes Transcription
Case No. I9603073

1140	Battalion Two:	Battalion Two is on the scene. Smoke showing from the roof.
1140	Dispatch:	Okay, Chief, and noted at 1140.
1140	Engine Three:	Engine Three to Battalion Two.
1140	Battalion Two:	Battalion Two.
1140	Dispatch:	Go ahead Chief.
1141	Battalion Two:	Strike me a second alarm.
1141	Dispatch:	Copy that.
1141	Battalion Two:	Flames showing.
1141	Dispatch:	Copy and noted.
1141	Battalion Two:	Get me some P.D down here to help me get these buildings evacuated.
1141	Dispatch:	Copy that.
1141	Eng---	Eng---(Tone.) (Tone.)
1141	Dispatch:	Zone Box 301-Advanced Auto, Indian River and Sparrow, Engine 14 and Engine Two start responding. Engine 14 and Engine Two respond for second alarm.
1142	Engine Two:	Engine Two responding
1142	Dispatch:	Okay, Engine Two.
1142	Eng:	Eng----
1142	Battalion One:	Battalion One to Battalion Two.
1142	Dispatch:	Battalion Two.
1142	Battalion Three:	Battalion Three is responding.
1142	Battalion One:	?????
1142	Dispatch:	Okay, Chief Three.
1142	Engine Three:	Engine Three to Battalion Two.

1142	Battalion Two:	Go ahead.
1142	Engine Three:	Chief, I'll need a crew bottle up insided. We got fire in ceiling.
1142	Battalion Two:	That's affirmative.
1142	Battalion Two:	We got a fire showing – – Building fire showing from the outside.
1142	Eng-	Eng? Responding.
1142	Engine Three:	Crew coming in; bring pike pole.
1142	Battalion One:	Battalion One to Battalion Two.
1143	Engine 14:	Engine 14 responding.
1143	Dispatch:	Okay.
1143	Engine Three:	Engine Three to Battalion Two.
1143	Battalion Two:	Battalion Two. Go ahead.
1143	Engine Three:	Chief, I'm gonna' need the second due engine to lay me supply line. I'm located in the rear of the building.
1143	Battalion Two:	Okay, Engine One, did you receive that?
1143	Engine One:	Engine One on the scene.
1143	Dispatch:	Okay.
1144	Engine Three:	Engine Three to Battalion Two.
1144	Engine Three:	Engine Three to Battalion Two.
1144	Dispatch:	Headquarters to Battalion Chief Two.
1144	Ladder Two:	Ladder Two is on the scene.
1144	Dispatch:	Okay Ladder Two.
1144	Engine Four:	Engine Four.
1144	Dispatch:	Engine Four.

Page 7
Dispatch Tapes Transcription
Case No. I9603073

1144	Engine Four:	Engine Four is back in service. (beep)
1144	Dispatch:	Okay. (beep, beep, beep, etc.)
1144	Battalion Two:	Battalion Two.
1144	Dispatch:	Go ahead Chief, Engine Three is trying to reach you.
1145	Battalion Chief:	Uhp'.
1145	Battalion Three:	Battalion Three.
1145	Dispatch:	Go ahead Chief.
1145	Battalion Three:	Go ahead and put Norfolk in Station One, and I'll advise you about Three, as soon as we get some more information.
1145	Dispatch:	Copy that Norfolk Station One.
1145	Engine Four:	Engine Four to Engine 42.
1145	Engine 42:	Go ahead.
1145	Engine Four:	Lieutenant, you heading toward Three?
1145	Engine 42	That's affirmative.
1145	Engine Four:	Okay.
1145	Battalion Two:	Battalion Two.
1145	Dispatch:	Go ahead Chief.
1146	Battalion Two:	I want a next due company to come into the front; I want to lay a line up to Advanced Auto.
1146	Engine One:	Engine One to Battalion Two.
1146	Engine One:	Engine One to Battalion Two.
1147	Battalion Two:	Battalion Two.
1147	Dispatch:	Go ahead Chief.
1147	Battalion Two:	Do you have another Battalion Chief?

Page 8
Dispatch Tapes Transcription
Case No. I9603073

1147	Dispatch:	Affirmative. Chief Harris is in route.
1147	Battalion Two:	You may as well notify Chief Elliot, Chief Best that we have a working commercial building fully involved.
1147	Dispatch:	Copy. Sir, will be notified.
1148	Engine Three:	Engine Three to Battalion Two.
1148	Engine Three:	Engine Three to Battalion Two.
1148	Dispatch:	Battalion Two.
1148	Battalion Two:	Battalion Two.
1148	Dispatch:	Were you trying to reach headquarters sir?
1148	Battalion Two:	Battalion Two.
1148	Dispatch:	Go ahead Chief.
1148	Battalion Two.	Might as well go on and strike me a third alarm. Staging on all sides. We need some staff support in here.
1148	Dispatch:	Copy sir. Also, be notified that we have one ladder company in the entire city. Tower 5 and Ladder 12 are both down.
1148	Battalion Three:	Battalion Three.
1148	Dispatch:	Battalion Three.
1148	Battalion Three:	Go ahead, send a ladder from Norfolk.
1148	Dispatch:	Copy that.
1148	Battalion Three:	You can get it at uh, Station 8 in Berkley.
1148	Dispatch:	Okay. (sirens in the background)
1149	Engine 42:	Engine 42 to Battalion Three.

Page 9
Dispatch Tapes Transcription
Case No. I9603073

1149	Battalion Three:	Go ahead 42.
1149	Engine 42:	Do you want us to respond to the fire scene?
1149	Battalion Three:	Yeah, he called for a third alarm. Go ahead. (Breaking up.)
1149	Medic One:	Medic One.
1149	Dispatch:	Medic One.
1149	Medic One:	Show us the route to this fire.
1149	Dispatch:	Copy Medic One.
1149	Battalion One:	Battalion One will be responding from the Chief's Office.
1149	Dispatch:	Okay.
1149	Engine 14:	Engine 14 to Battalion Two. We're coming up at Indian River Road now. You want to give us an assignment.
1149	Battalion Two:	_____
1149	??????????	Hello.
1149	??????????	????????????????????
1149	Engine One:	Engine One to Battalion Two.
1149	Battalion Two:	Battalion Two. Go ahead.
1149	??????????	????????????????????
1149	Engine Three:	Chief, I got Frank and myself and we can't get out.
1149	Battalion Two:	I can't understand you.
1149	Engine Three:	Chief, somebody to the front and get us out of here.
1150	Battalion Two:	Can anybody on this scene, on this fire? Can you hear or able to understand anything—transmission?

Page 10
Dispatch Tapes Transcription
Case No. I9603073

1150	Engine Four:	Unable to copy.
1150	Engine 14:	Engine 14 on the scene.
1150	Battalion Two:	Engine 14 go to the front of the building, find your hydrant to the front, lay to this ladder setting up.
1150	Engine 14:	We received, we'll catch this hydrant right out front here, and go to the front.
1150	Battalion One:	Battalion One to Battalion Two. Transmission sounds like it's somebody trapped inside of the building (phone ringing while this is going on).
1150	Battalion Three:	Battalion Three to Battalion one. I'll be there in just a minute to help us out.
1150	Battalion One:	Okay, pass on that transmission, sounds like somebody was trapped inside. Battalion Three, that's what I got. Let me get there, and I'll help.
1150	(Tone.)	
1151	Dispatch:	Engine 42 and Engine Four. Start responding to Indian River and Sparrow. Engine 42 and Engine Four respond to Sparrow and assist.
1151	Engine Four:	Engine Four is responding from Great Bridge Blvd.—No--
1151	Chief Four:	...Also Dispatch—Do you have Medic?
1151	Dispatch:	I have Medic One in route. Do you wish another?
1151	Chief Four:	Negative.
1151	EMS Three:	Medic Three's in route. Go ahead and put myself and Medic Three on the call.
1151	Dispatch:	Okay, Medic Three.
1151	Battalion Three:	Battalion Three. Medic Three responding.

Page 11
Dispatch Tapes Transcription
Case No. I9603073

1151	Medic One:	Medic One to EMS Three. Medic One is on the scene. Engine catching the hydrant in this middle of Indian River Road.
1151	Battalion Three:	Battalion Three.
1151	Dispatch:	Battalion Three.
1151	Battalion Three:	We'll need Ladder 10 out of Virginia Beach. They are about three blocks from this location. Ladder 10 out of Virginia Beach.
1151	Dispatch:	Copy that.
1151	Engine Three:	Engine Three to Battalion Two.
1151	Battalion Two:	Go ahead.
1151	Engine Three:	Captain Meads crew has been converted interior attack, to try and hold the fire in check.
1152	Engine Three:	Engine Three to Engine One. I need – water – as soon as you can get it to me.
1152	Battalion Two:	Okay, we've got it fully involved on the front of this building, I've got no water whatsoever up in the front.
1152	Engine 14:	Engine 14 is laying to the front right now.
1152	Medic One:	Medic One's on the scene.
1152	Medic One:	Medic One to dispatcher, we're on-scene at the hydrant.
1152	?????????	Be at the rear of the building.
1152	Engine 14:	Fourteen is on the scene, laying a 5".
	Medic Three:	Medic Three is on the scene.
1152	?????:	(Broken up transmission.)
1152	Battalion Three:	Battalion Three on the scene. All companies responding, switch to Channel 2.

Page 12
Dispatch Tapes Transcription
Case No. I9603073

1153	Captain Meads:	Captain Meads to Battalion Two.
1153	Portable 14:	Portable 14 to Engine 14. Now we're ready to charge the hydrant on your ready.
1153	Engine 14:	Engine 14 to Lt. Tarkington.
1153	Battalion Three:	Go ahead, start switching to Channel 2.
1153	Engine 14:	Engine 14 to Lt. Tarkington.
1153	????????????	Go ahead. Give us some water.
1153	???????????	Charge it. (Desperation sounds.)
1154	??????????	Unit calling to Lt. Tarkington. Are you ready for water?
1154	Battalion Three:	Battalion Three.
1154	Dispatch:	Go ahead Three.
1154	Battalion Three:	Go ahead tell that Ladder Company coming from Virginia Beach to set-up in front of BeLo's, and tell that Engine Company to lay their own supply lines.
1154	Dispatch:	Copy that. You want the ladder from Virginia Beach, and you an Engine from Virginia Beach as well?
1154	Battalion Three:	Correct, tell them to set-up right in front of BeLo's.
1154	Battalion Two:	Chief Harris, where are you?
1154	Battalion Three:	I'm in the front Chief, just getting into the now.
1154	Battalion Two:	All right, you come on down here to the other side, the foot, that's where I'm sitting.
1154	Battalion Three:	Okay, I'll be right there.

Page 13
Dispatch Tapes Transcription
Case No. I9603073

1154	EMS Three:	EMS Three to Dispatch. When you get a chance, bring a PSO out here. I got people running over this hose lines.
1154	Dispatch:	Copy that.
1154	EMS Three:	This lane is for emergency traffic.
1155	Engine One:	Engine One to Battalion Two.
1155	Engine 42:	Engine 42 to Battalion Three. Battalion Two.
1155	???????????	(Broke up.)
1155	Engine 42:	Chief, we're at the rear of Engine Three, we haven't had water, and the fire is getting ready to cook the engine.
1155	Battalion Two:	What unit is this?
1155	Captain Meads	We're sitting back here with dead lines and the fire is blowing right over us.
1155	Battalion Two:	You're in the back?
1155	Captain Meads:	I'm saying the fire is blowing directly over the engine, we're having to move it, and the lines are not effective.
1155	Battalion Two:	All right, you have your water?
1155	Engine 14:	Engine 14 to Lt. Tarkington.
1155	Captain Meads:	To Battalion Two.
1155	Battalion Two:	Battalion Two.
1156	Captain Meads:	I got men on the inside from Engine Three, and the lines have been burned; I do not know there status, and we still have water to go in after them.
1156	Engine 42:	Engine 42 to Battalion Three.
1156	Battalion Three:	Battalion Three.

APPENDIX C

Changes Initiated by the CFD Since the Fire
Chesapeake Fire Department Initiatives
Truss Identification Program

CHESAPEAKE FIRE DEPARTMENT INITIATIVES

OPERATIONS

C   Automatic response of 2nd BC on any working incident to support command structure.

C   Developed policy of IDLH atmospheres, including on-site Rapid Intervention Team before entry of interior crews.

C   Automatic response of EMS Captain to support command staff with personnel accountability.

C   Automatic switching to Channel 2 (Fireground Tactical Channel) upon arrival on the scene.

C   Developed comprehensive Operations Policy concerning communications and the use of radio equipment.

C   Developed Truss Identification (TIP) to identify and mark all truss construction within the City. Information is included in database and pre-plan books.

C   Reinforce use of existing Incident Command System (ICS).

C   Reinforce use of existing Personnel Accountability System (PAS).

C   Reinforce use of existing PASS device use.

C   Training staff automatically deployed to all working incidents, 2nd alarm or greater, to function as Safety Support Officers.

C   Formed Research Committee to revise current pre-plan documentation data using computer based drafting, mapping, and storage capabilities.

EQUIPMENT

C   Enhanced ICS Command pack with additional resources to improve command capabilities.

C   Added communication equipment to ICS Command Pack to provide ability to communicate/ monitor both fire channels simultaneously.

C   Added additional portable radios to each piece of apparatus.

C   Research and development program to compare all current S.C.B.A. equipment on the market, and make recommendations to the department.

TRAINING

P   Mandatory State certification for Incident Safety Officer for all Battalion Chiefs and Staff Chiefs.

C = Completed

P = Planned next year

City of Chesapeake Fire Department	Standard Operating Procedure		
	No. 50.20	Date	10/09/96
	Subject Truss Identification Program (TIP) Page 1 of 3		

I.   PURPOSE

The use of trusses in building construction presents a great danger to firefighting personnel when those structures are involved in fire conditions. By design, the truss members in floor and roof assemblies will collapse, without warning, after being exposed to heat or flame contact for a very short period of time. Because of the inherent danger firefighters must face while operating within these buildings, a Truss Identification Program (TIP) has been instituted to alert personnel of the danger prior to beginning fire suppression operations. The Truss Identification Program is intended to alert the members of the Chesapeake Fire Department with pertinent pre-plan information before firefighting forces are committed to an interior attack.

II.   SCOPE

The following guidelines shall be followed whenever a truss assembly is discovered in any commercial building in the City of Chesapeake.

III.   POLICY

The TIP shall be an ongoing program applied to all commercial buildings inspected by the Chesapeake Fire Department. Overall responsibility for implementation and control of the program will be under the direction of the Deputy Chief of Operations. Station Commanders will insure that all commercial buildings within their district are inspected in accordance with the City of Chesapeake, Virginia Fire Inspections Manual.

Company Officers will insure that personnel under their command are trained, and understand different types of building construction and truss assemblies.

City of Chesapeake Fire Department	Standard Operating Procedure		
	No. 50.20	Date	10/09/96
	Subject		
	Truss Identification Program (TIP)		
	Page 2 of 3		

IV.   PROCEDURE

A. NOTIFICATION

1. Upon discovering a truss assembly, fill out a <u>Truss Identification Program Reporting Form</u> and forward it to the Information Management Office.

2. When the <u>Truss Identification Program Reporting Form</u> has been received by Information Management, truss information will be entered on the C.A.D. system premise file. The ringdown printout sheet received by stations on the initial alarm will alert companies of the use of Truss Construction in the structure to which they are responding.

B. PRE-FIRE PLAN TRUSS LABELS

1. One florescent orange Pre-fire Plan Truss Warning Label shall be applied to the top center portion of the pre-fire plan when truss assemblies are identified as part of the construction members of that particular building. While enroute to the scene, pre-fire plans should be placed on the accountability clipboard for easier access and review by arriving Incident Commanders.

C. APPLYING TRUSS IDENTIFICATION STICKERS ("T" STICKERS)

1. Explain to the business owner that "T" stickers do not mean that their building is hazardous. It does not affect their insurance rates or cause any other concern for the business owner. They are only applied to alert firefighters of a potential danger of collapse of the building under fire conditions. In all cases, Fire Department personnel should be polite, courteous, and professional in their actions. A fact sheet handout should be left with the owner or representative of each building where a "T" sticker is applied.

City of Chesapeake Fire Department	Standard Operating Procedure		
	No. 50.20	**Date**	10/09/96
	Subject **Truss Identification Program (TIP)** **Page 3 of 3**		

2. If the business owner still object to the "T" sticker, comply with his/her wishes and refer the matter to the Fire Marshal's Office. Enforcement will be in accordance with Chesapeake City Code Section 34-3.

3. Upon obtaining the business owner's permission to apply the "T" sticker, use the following procedure.

   a Thoroughly clean an area on the top right-hand corner of the main entry door.

   b Apply the pressure sensitive "T" sticker.

   c In multiple occupancy structures (shopping center, strip malls, etc.) it will be necessary to place a "T" sticker at each occupancy address.

   d In cases where the Fire Department may make initial entry through the rear door of an occupancy, a "T" sticker should also be applied to that door.

   e In cases where one business occupies several addresses, your own judgment must be used as to the number of "T" stickers applied.

   f In cases of multiple adjacent doors, only the center-most door should be marked. It is not necessary to mark every door when several doors are located at a common entrance point to a structure.

   g Judgment should be exercised so that existing lettering or advertisement is not covered up or disturbed.

APPROVED: _Michael O'Brien_

APPENDIX D

Photographs of the Advance Auto Fire

An elevated view of the rear of the store, showing (1) the door in the rear where the firefighters entered with the 1 3/4 attack line; (2) the trailer that blocked the view of the rear entrance; (3) the location of the utility pole; and (4) the employee bathrooms where the fire was originally discovered. (*Photo courtesy of the NFPA*).

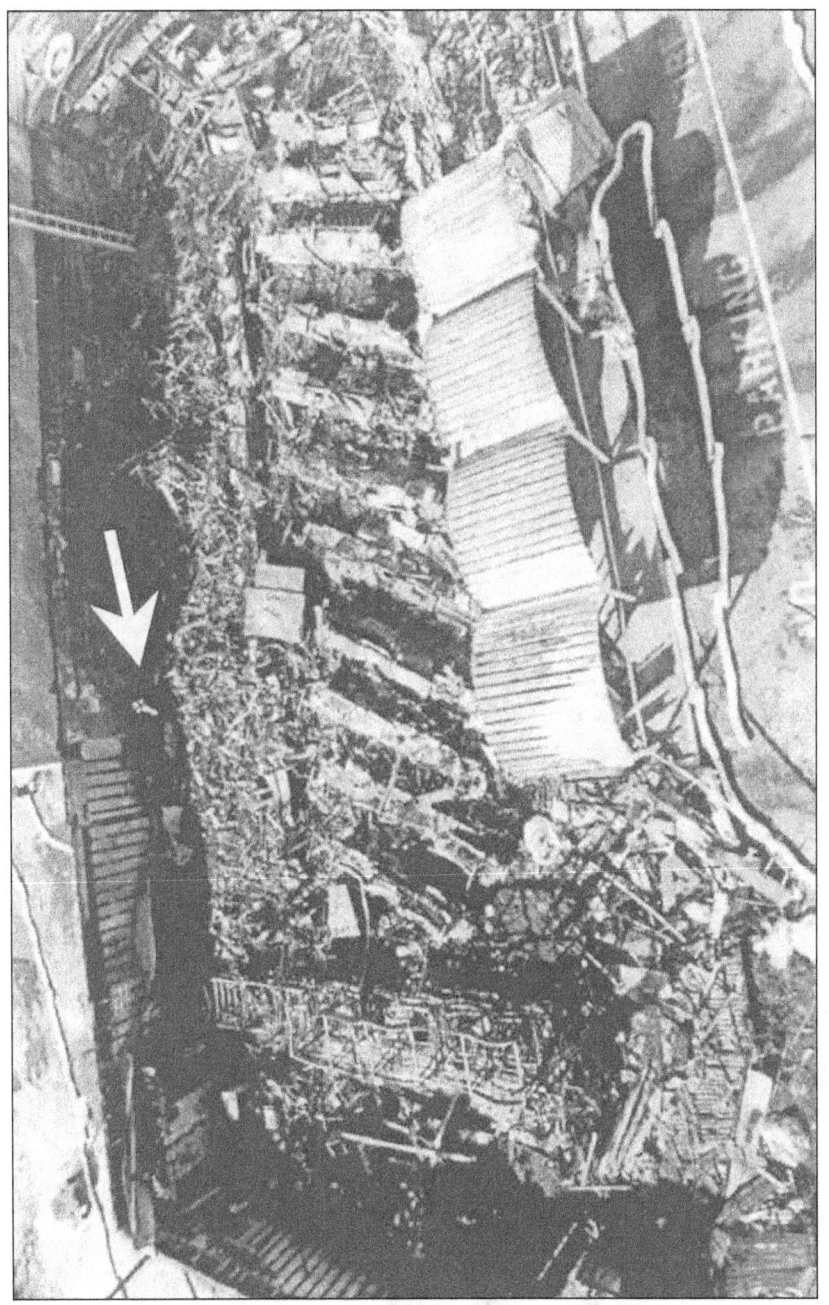

Photo of interior of the store from an elevated position on side two. The white cross (arrow) indicates where both firefighters were found. *(Photo courtesy of the NFPA).*

The truss loft area in another part of the Indian River Shopping Center. (Photo courtesy of NFPA).

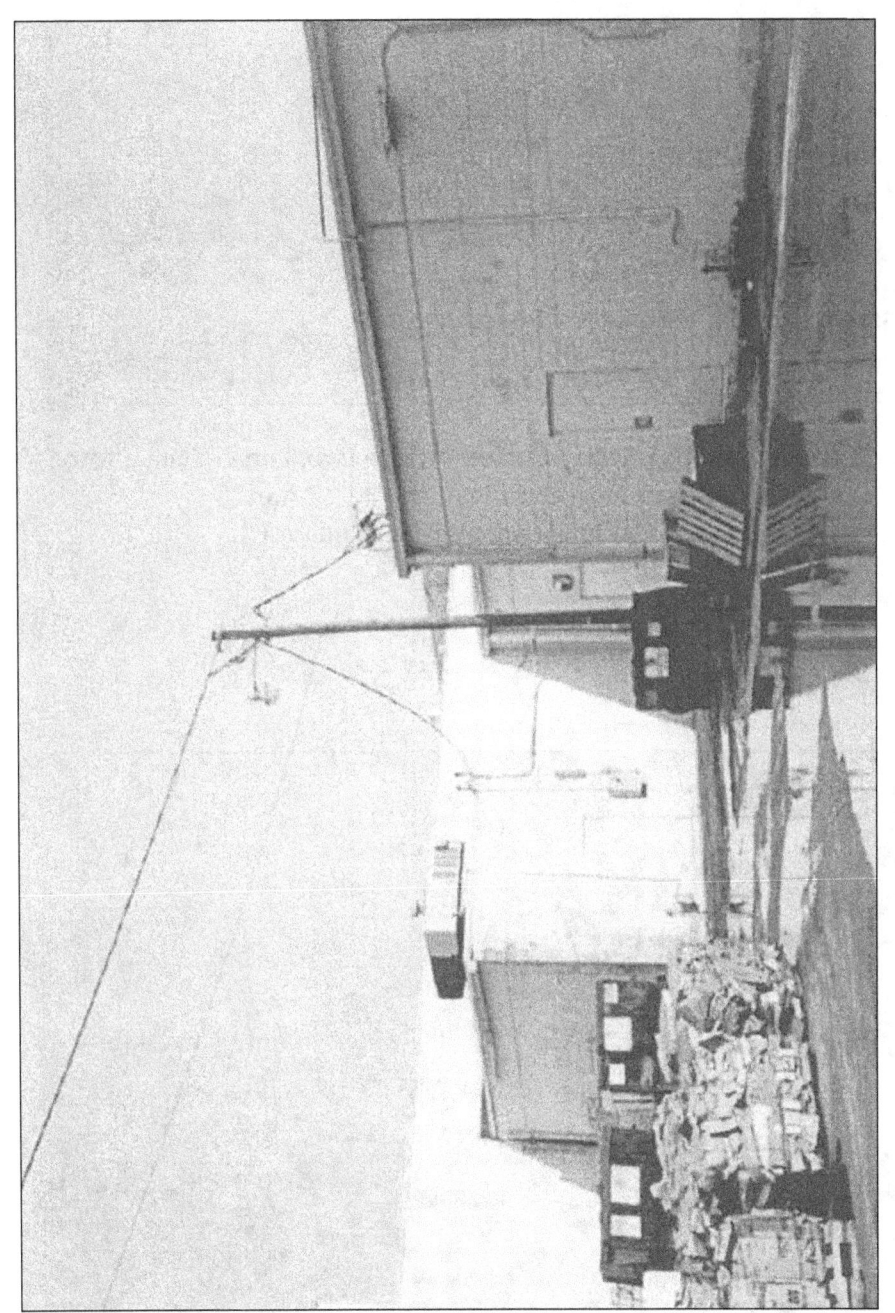

Utility pole connecting power to a meter at the rear of the Indian River Shopping Center. (USFA file photo).

Ladder 2 is shown in its initial position on side two. Time of the photo
is estimated between 11:50 and 11:52 a.m.
(*Photo courtesy of Chesapeake Fire Department*).

Ladder 2 is shown in position on side one. Time of the photo is estimated
at 11:52 a.m. The supply line in the front of the picture is from Engine 1.
(*Photo courtesy Chesapeake Fire Department*).